# Indigenous-Settler Relations in Australia and the World

Volume 4

The series, Indigenous-Settler Relations in Australia and the World, brings together scholars interested in examining contemporary Indigenous affairs through questions of relationality. This is a unique approach that represents a deliberate move away from both settler-colonial studies, which examines historical and present impacts of settler states upon Indigenous peoples, and from postcolonial and decolonial scholarship, which is predominantly interested in how Indigenous peoples speak back to the settler state. Closely connected to, but with meaningful contrast to these approaches, the Indigenous-Settler Relations series focuses sharply upon questions about what informs, shapes and gives social, legal and political life to relations between Indigenous peoples and non-Indigenous peoples, both in Australia and globally.

This is an important and timely endeavour. In Australia, relations between Aboriginal and Torres Strait Islander peoples and the state are at an impasse. In the wake of the government's rejection of the Uluru Statement in 2017 there is no shared view on how Indigenous-settler relationships might be 'reset', or even if this is possible. The contemporary Indigenous affairs policy domain is characterised by confusion, frustration and disappointment that, despite a seemingly endless succession of policy regimes, efforts to 'close the gap' between Aboriginal and Torres Strait Islander peoples and other Australians have not resulted in progress.

It is into this contested space that the Indigenous-Settler Relations series seeks to intervene with new, agenda-setting research. The series editors are based in a research unit in the Faculty of Arts at the University of Melbourne—the Indigenous Settler Relations Collaboration. The series will build on the work of the Collaboration in bringing together scholars and practitioners from around Australia, and around the world—particularly other Anglophone settler colonial societies such as Canada, the United States and New Zealand—whose work is concerned with Indigenous-settler relations across a range of disciplines. The multi-faceted approach to Indigenous-Settler Relations that defines the series seeks to capture how the question of relationality is already being asked by scholars across disciplines including political science, history, sociology, law, media, and cultural studies.

Readers of this series will look to it for fresh perspectives and new ideas about how to transform Indigenous-settler relations in Australia and elsewhere. They will learn from the leading lights in an emerging field who will connect their rich, multi-disciplinary scholarship to urgent social and political questions at the heart of Indigenous-Settler relations.

Nikki Moodie · Sarah Maddison
Editors

# Public Policy and Indigenous Futures

Springer

*Editors*
Nikki Moodie
School of Social and Political Sciences
University of Melbourne
Parkville, VIC, Australia

Sarah Maddison
School of Social and Political Sciences
University of Melbourne
Parkville, VIC, Australia

ISSN 2524-5767 ISSN 2524-5775 (electronic)
Indigenous-Settler Relations in Australia and the World
ISBN 978-981-19-9318-3 ISBN 978-981-19-9319-0 (eBook)
https://doi.org/10.1007/978-981-19-9319-0



This Springer imprint is published by the registered company Springer Nature Singapore Pte Ltd.
The registered company address is: 152 Beach Road, #21-01/04 Gateway East, Singapore 189721, Singapore

# Acknowledgements

We wish to acknowledge that many individuals and communities have made this book possible. Much of this work has been delayed by the COVID-19 pandemic, and we are especially appreciative if the patience shown by many of the people involved. Many of the chapters in this collection were developed at a research workshop funded by the Australian Political Studies Association (APSA), originally planned for 2020 and eventually held late in 2021. In particular we'd like to thank Elizabeth Strakosch, who contributed significantly to the development of the volume.

Support for this work also came from the staff in the Australian Centre at the University of Melbourne. Originally established in 1989, the Australian Centre has a long history of supporting scholarship of colonial and contemporary Australian society and culture. Since the beginning of 2022, the research programs of the Centre have been revitalised under the directorship of Professor Sarah Maddison and deputy directorship of Dr. Julia Hurst. The Australian Centre advances critical understandings of how Australia's founding as a settler colony continues to inform, shape, and constrain our capacity to engage with the central challenges of our age, while simultaneously recognising that "Australia" is not the first, the only, nor the central, political organising principle of these lands and waters. The Australian Centre is committed to research that works towards more just relations between Indigenous and settler peoples, and currently supports the publication of the Indigenous-Settler Relations in Australia and the World series.

The Australian Centre's Research Coordinator, Eleanor Benson, kept all of us on track during the process of developing this book, including organising the workshop in 2021. Eleanor also provided outstanding editorial and proof reading assistance in the final stages of completing this manuscript and we are deeply grateful to her for the always exceptional quality of her work.

Most importantly, this book has been made possible due to the Wurundjeri peoples of the Kulin nations upon whose unceded territories the Parkville campus of The University of Melbourne resides. Our occupation of their land is an injustice that will possibly never be made right. These words alone do very little to provide deserving reparations for that irreparable loss. We sit with that discomfort daily while working towards better relations between our institutions and First Nations peoples everywhere.

# Contents

# Editors and Contributors

## About the Editors

**Nikki Moodie** is the current Program Director and Deputy Director of the Atlantic Fellows for Social Equity at the University of Melbourne. Nikki is an Associate Professor of Sociology and a queer Gamilaroi woman. Nikki holds a Bachelor of Arts with Honours in Political Science from the University of Queensland, and a Ph.D. in Sociology from the Australian National University. Nikki's research focuses on Indigenous education and social policy.

**Sarah Maddison** is a Director of the Australian Centre and Professor of Politics at the University of Melbourne. With Associate Professor Sana Nakata and Dr. Julia Hurst she edits the Springer book series Indigenous-Settler Relations in Australia and the World. She has published widely in the fields of reconciliation, settler colonialism, and Indigenous politics. Her most recent book, The Colonial Fantasy, was published by Allen & Unwin in 2019.

## Contributors

**Morgan Brigg** The University of Queensland, St Lucia, QLD, Australia

**Anthea Compton** Jumbunna Institute for Indigenous Education and Research, University of Technology Sydney, Ultimo, NSW, Australia

**David Coombs** School of Education, University of New South Wales, Kensington, NSW, Australia;
The Australia and New Zealand School of Government (ANZSOG), Carlton, VIC, Australia

**Debra Evans** Charles Sturt University, Wagga Wagga, NSW, Australia

**Janine Gertz** Nura Gili, University of New South Wales, Kensington, NSW, Australia

**Mary Graham** The University of Queensland, St Lucia, QLD, Australia

**Miriam Jorgensen** Native Nations Institute, University of Arizona, Tucson, AZ, USA

**Sarah Maddison** The Australian Centre, University of Melbourne, Parkville, VIC, Australia

**Justin McCaul** Australian National University, Canberra, ACT, Australia

**Nikki Moodie** The University of Melbourne, Parkville, VIC, Australia

**Donna Murray** University of Technology Sydney, Ultimo, NSW, Australia

**Will Sanders** Australian National University, Canberra, ACT, Australia

**Anya Thomas** The Australian Centre, University of Melbourne, Parkville, VIC, Australia

**Alison Vivian** Jumbunna Institute for Indigenous Education and Research, University of Technology Sydney, Ultimo, NSW, Australia

# Chapter 1
# Introduction: Public Policy and Indigenous Futures

**Nikki Moodie and Sarah Maddison**

**Abstract** This volume is driven by a question over whether public policy, rendered as 'Indigenous affairs', can be motivated by Indigenous futurity, rather than the eliminatory desire of settler colonialism. The project of securing a future for settlers is always predicated on the replacement of Indigenous peoples and the theft of Indigenous land; the reality that 'Indigenous affairs' largely does not elevate Indigenous rights and wellbeing suggests that the settler state is in fact achieving what it sets out to do. 'Indigenous affairs'—as a mode of colonial governance—problematises the Indigenous subject in order to incapacitate Indigenous collectivities. The history of state 'failure'—of targets missed, of underfunding, of violence, of racism, of precarity—does not manifest as a surprising and unintended consequence of colonial policy. Political contests are fought over the future as much as they are determined by the past, and so the practice of failure simply moves the desired state of Indigenous vanishment back in time to the present. A particular version of an Indigenous future is always present in settler colonial Indigenous policy, and this is and always has been fought by Indigenous peoples who have a different relationship with both the past and the present.

**Keywords** Social policy · Governance · Decolonization · Self-determination · Closing the gap

## 1.1 Introduction

'Failure' is a persistent narrative in Australian Indigenous affairs, and government approaches to Indigenous issues are widely recognised as falling short of the limited objectives those governments have set for themselves: to 'close the gap' on a range of statistical indicators between Aboriginal and Torres Strait Islander people and the wider population (Aitken, 2009; Anderson, 2015; Strakosch, 2019). In efforts to achieve this aim, and in earlier iterations of policy ambition, government approaches

N. Moodie (✉) · S. Maddison
The University of Melbourne, Parkville, Australia
e-mail: nikki.moodie@unimelb.edu.au

 
N. Moodie and S. Maddison (eds.), *Public Policy and Indigenous Futures*,
Indigenous-Settler Relations in Australia and the World 4,
https://doi.org/10.1007/978-981-19-9319-0_1

to Australian Indigenous affairs have undergone endless overhaul. Regimes of protection, assimilation, self-determination, and intervention have been advanced and withdrawn by ever-changing colonial structures. During the last 50 years there have been more than 20 different ministers in the federal government Indigenous affairs portfolio, and ten different organisational structures, nine of which have been created/dismantled within the past 30 years (Patterson, 2017, pp. 15–16). For Aboriginal and Torres Strait Islander people and communities, the experience of constant federal policy change is frustrating, destabilising, and disempowering, underscoring settler society's disregard for Indigenous sovereignties and autonomy. The constant turmoil in the states' approach to Indigenous affairs lends weight to Patrick Wolfe's observation that settler colonial societies direct a disproportionate amount of energy and effort towards administering the lives of Indigenous peoples, who are generally very statistically small groups in the society (Wolfe, 2006, pp. 37–8).

Yet the failure of settler policy is not the only, or even the most important, dynamic in Indigenous affairs. As the settler state flounders in its quest to deliver 'good policy', Indigenous peoples, communities, organisations, and scholars in Australia are engaged in projects that Indigenous scholars have described as acts of resurgence and refusal (see for example Carlson & Frazer, 2021; Grande & McCarty, 2018; Simpson, 2017a, 2017b; Vizenor, 2008). Rather than looking to the settler state for policy solutions, refusal and resurgence see Indigenous peoples turn away from settler institutions, values, and ethics and towards their own institutions, values, and ethics through modes of self-determination and self-governance (Begay et al., 2007). The survival and resurgence of Indigenous governing and policy expertise, which speaks to the resilience of cultures and communities globally, is offering new agendas and pathways towards policy that is developed and implemented by Indigenous peoples (Smith et al., 2021). The focus on Indigenous resurgence provides a break with settler narratives of crisis and deficit that have long described Indigenous affairs in Australia particularly. Beyond resilience and resurgence, we see vitality and efflorescence—futures that are challenged by ongoing colonisation, but which are driven and given voice by the hope, labour, and persistence of Indigenous peoples (Roche et al., 2018).

As the contributors to this volume met together for online writing sessions (navigating the impacts of the COVID-19 pandemic), we drew on the work of scholars oriented towards resurgence and refusal (e.g. Alfred & Corntassel, 2005; Simpson, 2016) and rebuilding native nations (e.g. Cornell, 2015; Jorgensen, 2007), to help us make sense of this critical disjuncture in thinking about 'Indigenous policy'. From these conversations we together and individually come to consider the complexity of 'Indigenous policy' as the exercise of state power over Indigenous issues very broadly, and a different rendering of Indigenous policy as those strategic objectives determined—and in turn implemented—by and for Indigenous peoples. We hope these chapters mark what may be a turning point in the governance of Indigenous lives and futures in Australia. As the settler state clings to its assumed legitimacy at the political and policy interface with Indigenous peoples, First Nations and Indigenous organisations are increasingly going their own way, exercising their rights to self-determination, and rejecting the suggestion that they require recognition from the state in order to (re)build their nations and autonomous governing capacity.

## 1.2 The Dawn of a New Era?

The Australian settler state is certainly eager to suggest that it is entering a new era in Indigenous affairs. At both national and sub-national level, governments are opening up new spaces and modes of engaging with First Nations' governance. The emerging treaty processes in Victoria, Queensland, and the Northern Territory are one such shift in engagement, the possibilities of which remain a long way from being realised but are also multiple and reparative. The National Agreement on Closing the Gap also represents itself as a 'a fundamentally new way of developing and implementing policies and programs that impact on the lives of Aboriginal and Torres Strait Islander people' (Closing the Gap 2020). So too does the co-designed proposal for an Indigenous 'Voice to Government', which suggests it 'will be part of genuine shared decision-making with governments at the local and regional level and have our voices heard by the Australian Parliament and Government in policy and law making'. This, the proposal suggests, could be 'the most significant reform in Aboriginal and Torres Strait Islander affairs for generations' (Commonwealth of Australia, 2021, pp. 6–7). This process seeks to establish the Voice to Government as an alternative to the constitutionally enshrined Voice to Parliament that was proposed in the Uluru Statement from the Heart—a set of proposals we discuss further below.

At the same time as we observe the possibility that a new era may be emerging, however, we also recognise that—as ever—the settler state cannot be trusted to support greater Indigenous autonomy and self-determination. Settler colonial logics of elimination (and the desire for colonial completion) have always undercut and undermined such moments of apparent transformation in Indigenous-settler relations (Maddison, 2019). Thus, even as these new doors to more genuine partnerships and First Nations' autonomy are fought open, we also see a resurgence of First Nations and Indigenous organisations seeking to decouple from their relationships with the state. In the chapters by David Coombs and Justin McCaul, for example, we see the depth of Indigenous policy and administrative expertise in sectors constrained by settler colonial state governance, and in their chapter Miriam Jorgensen and colleagues describe both the benefits and urgency of adopting a nation-building approach to enable the continued flourishing of Indigenous lives. Such approaches, the authors note, need no permissions to proceed.

Of course, this rejection of settler authority has always been a part of Indigenous political culture on this continent. Indigenous resistance to the colonial order has been present from the first days of invasion and has been sustained in the centuries since. But during the second half of the twentieth century in particular, there were also notable efforts by Indigenous people and organisations to work *with* the state through structures such as the Aboriginal and Torres Strait Islander Commission (1990–2005), or in creative and pragmatic ways simultaneously with and against government policy and funding arrangements (see the contributions in Maddison & Brigg, 2011, *inter alia*). That creativity and pragmatism is evident we hope in this volume. This is important policy work that seeks to crack the stone façade of the colonial order and create space for Aboriginal and Torres Strait Islander peoples' own solutions to flourish.

Yet these efforts at creativity and pragmatic engagement with state policies and institutions have, by definition, operated within the parameters set by the state. Indeed, the settler state has been prepared to accommodate (even take credit for) many of these innovations in Indigenous governance. At every turn, however, the settler order has also resisted the push for more substantive and structural reform in the relationships between First Nations and the state. In the 50 years since the Aboriginal Tent Embassy was established at (now Old) Parliament House in Canberra, with its symbolic and material expression of abiding Indigenous sovereignties, the settler state has not moved an inch towards genuine engagements with First Nations *as nations*. Aboriginal and Torres Strait Islander peoples on this continent are still treated as one among many multicultural minorities rather than as sovereign nations with abiding connections to Country and independent systems of law. At every turn, the settler state has sought to contain, control, and manage First Nations, including through the churn of policy discussed above, which limits the capacity for First Nations to exercise their own self-determination and the state to genuinely progress treaty.

Indeed, calls for treaty have a long history in Australia but, aside from a brief period of commitment to a national treaty by the Hawke government in 1988 (subsequently reneged and with a ten-year reconciliation process offered as a consolation prize), such calls had been resisted. Prime Minister John Howard famously rejected the call for treaty with the argument that 'an undivided nation does not make a treaty with itself' (Laws & Howard, 2000); nearly two decades later Prime Minister Malcolm Turnbull would reject the Uluru Statement from the Heart on a similar basis.

The response to proposals contained in the Uluru Statement from the Heart—for a constitutionally enshrined Voice to Parliament and a Makarrata Commission to oversee a process of agreement making and truth-telling (often referred to the in the shorthand call for Voice, Treaty, Truth)—perfectly illustrate the limitations of the Australian Government's approach to Indigenous affairs. The Uluru Statement emerged from the work of the Referendum Council, which was appointed by then prime minister Malcolm Turnbull to lead community consultation on the nature of constitutional recognition (Anderson & Leibler, 2017). From December 2016 through 2017 the Referendum Council undertook a series of regional dialogues, which culminated at Uluru in the Northern Territory in May 2017. The powerful language of the Statement called for structural change in the relationship between Indigenous peoples and the state, and an incremental approach to treaty-making. The delivery of the statement, read by a lead member of the Referendum Council Professor Megan Davis, standing in the red dirt of Uluru, was a powerful moment in which—again—it seemed as though a new era in Indigenous-settler relations might be possible. Yet the statement was greeted with silence and eventually rejection by the government of the day. The proposal for gradual structural reform—as outlined in the Uluru Statement, with its Voice to Parliament working alongside the established political order to make Indigenous peoples' voices heard—seemed too much for the settler state. In October 2017, after months of silence, Prime Minister Turnbull rejected the Voice proposal claiming (incorrectly) that it 'would inevitably become seen as a third chamber of Parliament' (Turnbull et al., 2017).

Both Howard and Turnbull, of course, misrepresented Indigenous rights to self-determination and the possibilities that treaty, recognition, and reparation offer both Indigenous and non-Indigenous people. Government policy has acknowledged Indigeneity but not nationhood, and while successive governments claim to pursue distributional justice, they have failed to address demands for positional justice (Cornell & Jorgensen, 2019, 2020). Policies such as 'Closing the Gap' treat Indigenous peoples primarily as individuals in need of support rather than as nations with thousands of years of experience in managing sophisticated and grounded political systems, as Mary Graham and Morgan Brigg's chapter in this volume makes clear. Underlying the kinds of incorporated governance arrangements often demanded by the state in order for it to 'recognise' an Indigenous polity (the Victorian *Traditional Owners Settlement Act* 2010 is an example of framework that makes such demands), First Nations' governance survives, giving rise to policy expertise based on localised ways of knowing and being. Much of this expertise is focused on a future that looks very different to the oppressive and coercive policy regimes that characterise the settler order.

## 1.3 Policy and the Future

This volume is fundamentally driven by a question over whether public policy, rendered as Indigenous affairs, can be motivated by something other than its eliminatory desire. The project of securing a future for settler colonists is always predicated on the replacement of Indigenous peoples; Tuck and Gaztambide-Fernández (2013) suggests that any work that does not aim to interrupt this defining feature of settler colonialism is simply 'fettered to settler futures' (p. 80). The long history of failure in Indigenous affairs, when interpreted through this frame, therefore suggests that 'Indigenous policy' as a state project is achieving precisely what it has always set out to do. Indigenous affairs—as a mode of colonial governance—problematises the Indigenous subject as a strategy that is intended to incapacitate Indigenous collectivities. The long history of 'failure'—of targets missed, of underfunding, of violence, of racism, of precarity—does not manifest as a surprising and unintended consequence of benevolent colonial policy. Political contests are fought over the future as much as they are determined by the past, and so the practice of failure simply moves the desired state of Indigenous vanishment back in time to the present (Baldwin, 2012; Tuck & Gaztambide-Fernández, 2013). A particular version of Indigenous future is always present in settler colonial Indigenous policy, and this is and always has been fought by Indigenous peoples who have a different relationship with both the past and the present.

In this volume, the contributors are engaged in a range of projects that seek to move beyond the rhetoric of failure in Indigenous affairs. Central to this work is acknowledging that there are changes occurring at all levels of government whereby a different relationship between Indigenous polities and the settler colonial state seems possible. In tentative moves towards treaty or policy co-design, for example, state and federal governments betray a sense of the illegitimacy that defines their occupation of this continent and a growing recognition that a compact with Indigenous people, rendered as citizens of a liberal democracy, is an insufficient basis for a political relationship. Whilst the structure of that relationship is defined by conflict and struggle, and will likely always remain so, the more significant struggles are those that Indigenous collectivities themselves face to organise and act in the furtherance of their own aspirations (Cornell, 2015).

This collection highlights the kinds of self-determined and community-owned responses to complex socioeconomic and political changes that Indigenous peoples are driving, and the Indigenous-led policy development and policy expertise required to implement those changes. The chapters that follow highlight new Australian work in what is an international phenomenon (Smith et al., 2021) and focus on issues central to policy change and Indigenous futures. The chapters in this book also foreground the resurgence that is taking place in Indigenous governing and policymaking, providing case studies of local or community-based policy development and implementation.

Mary Graham and Morgan Brigg, in their chapter 'Indigenous Public Policy Futures: A Manifesto for Relationalist Public Administration', address the foundational illegitimacy of settler administration on the Australian continent through reorienting to a relational governance developed and refined by Indigenous people over tens of thousands of years. Settler sovereignty, they argue, cannot draw upon a longstanding tradition of administration or governing in this place and its attempted transplantation of European traditions remain youthful, insecure, and uncertain. The unhealed colonial wound affects all extant people and systems that seek to coexist with these lands; a political settlement that provides a future belonging and a new identity is possible if that settlement can be based on Indigenous philosophical principles. Graham and Brigg share an extended discussion of relationalism and its features, including the principles of continuity and interdependence, stewardship, and proportionality. They use three examples—Indigenous gender orders, a national Indigenous representative body that briefly existed from 2011 to 2017, and 'Caring for Country' programs—to demonstrate possibilities for legitimate authority in Australian public administration. These possibilities, which emerge from the process of embedding relational values—that are themselves foundational to Aboriginal political ordering—into Australian governing more broadly, can hence become shared foundations for a secure future of belonging and balance for newcomers and Indigenous peoples in this place. Graham and Brigg's approach to relational public administration creates the possibility of a more legitimate and mature leadership, based on older accountabilities than those which have more recently arrived.

The deep, relational approach to governance that Graham and Brigg articulate helps to make sense of the kinds of 'failure' that Will Sanders has observed over his long career as a scholar in this field. In his chapter 'Success and Failure in Australian Indigenous Policy: Moral Dynamics and Rhetorical Registers', Sanders offers reflections from some 40 years working in Indigenous affairs and Australian public policy. Focusing on the moral dynamics of governing and the rhetorical registers deployed by the state in Indigenous affairs, Sanders draws on Colebatch's and Bacchi's constructivist approaches to explore the normative dimensions of problem construction and representation. Identifying narratives of 'improvement' as central to Indigenous affairs, Sanders suggests that governments are motivated by a desire to lay claim to the future. Necessarily, this often requires the past to be disowned, despite the path dependency that delimits possible nominal 'successes' or 'failures' in Indigenous policy. Sanders explores three different policy phases and timeframes, showing how competing principles and alternative idioms emerged in the early part of the 21st Century to sediment the language of 'failure' in part as a challenge to improve government action over time. Sanders' chapter prompts us to consider the complex interplay of different visions of the future against deep knowledge of 'what works'. How the language of failure is deployed, often by governments themselves who seek to create a point of difference in their tenure, is one way that we can see claims to the future surfaced in Indigenous affairs.

Following this problematisation of settler time and what that means for Indigenous policy, Justin McCaul draws our attention to one example of Indigenous policy creativity. McCaul's chapter on 'Caring for Country as Deliberative Policymaking', describes how Indigenous expertise and leadership in land management is self-government manifest. Drawing on a decade of experience in Caring for Country programs and his recent doctoral research, McCaul first describes how the native title regime has not always led to substantive political power or sustainable economic benefit for Aboriginal and Torres Strait Islander communities. Despite its limitations, however, native title has made possible the legal recognition of a type of 'Indigenous estate', over which Indigenous-led land management can be exercised at least in part. McCaul notes that Aboriginal and Torres Strait Islander people now have some form of tenure or management responsibility for more than 20 per cent of the continent's land mass. In the generation since the High Court's Mabo decision, Caring for Country has become a vehicle for the assertion of rights, strengthening cultural institutions, decision-making, and financial security. As such, McCaul argues that even though native title and Caring for Country programs are ostensibly focused on land management, they provide an opportunity to exercise the collective decision-making that is central to Indigenous governing. But beyond this, when Aboriginal and Torres Strait Islander people ensure that land management practices are participatory and deliberative, they are modelling more democratic approaches to policy-making than the Australian state often allows. The work of Indigenous peoples in exercising self-government over land, whilst simultaneously and actively constituting more democratic processes in Australia, is a critically important insight offered by this chapter.

This type of creativity is also evident in Davis Coombs' chapter, 'Future-Proofing Indigenous Self-Determination in Health'. The complex strategies required by Indigenous organisations to maintain viability in the Australian settler colonial context are the subject of Coombs' work with Aboriginal Community Controlled Health Organisations (ACCHOs) in New South Wales. Coombs traces federal government opposition to self-determination through successive policies aimed at defunding

ACCHOs and increasing their exposure to market forces, and reports on 29 interviews at 18 Aboriginal health organisations across NSW that engage with these dynamics. Whilst the strategies adopted by those ACCHOs are innovative and pragmatic responses to an increasingly complex operating environment, the data shows a sense of risk to gains that have been made in Indigenous health. Coombs collates the evidence on the remarkable effectiveness of Aboriginal community-controlled health services, noting that the sector is the single largest employer of Indigenous people in the country and the role that these organisations play in addressing a wide range of the social determinants of health for Indigenous people and communities. Not only are the achievements of ACCHOs extraordinary in their impact on the lives of Indigenous people and communities, but these organisations also tend to be more effective and outperform their counterparts in the mainstream health sector. The adaptations that ACCHOs have been forced to make in response to increased marketisation, whilst innovative, have been made in the context of aggressive defunding and government opposition to self-determination in principle and practice. Coombs' chapter is a timely empirical analysis of Indigenous-led strategy for self-determination, and a reminder of the absolute necessity of Aboriginal control over Aboriginal health.

Moodie's chapter turns to self-determination in work and schooling, exploring the new education and employment targets in the latest iteration of the Australian Government's Closing The Gap policy. A persistent metaphor in these policy areas is the way that schooling, and low paying or insecure work is purported to act as a 'stepping stone' to more stable and higher paying jobs and better life chances in adulthood. Yet such approaches neglect the economic deregulation and dislocation that has made precarity and insecurity a feature of contemporary adulthood broadly, and the ways in which Indigenous peoples' pathways through education and work have been made even more precarious by structural and interpersonal racism specifically. Moodie suggests that at least part of Indigenous peoples' experiences of precarity is explained by the absence of self-determined stepping stones; where Indigenous lives are supported by Indigenous institutions, Indigenous outcomes improve. Whilst the refreshed CTG targets do include greater recognition of Indigenous lifeworlds and the right to determine our own futures and aspirations, the 'stepping stones' currently on offer are neither easy to walk nor lead to self-determined Indigenous futures.

Yet, perhaps, the state will remain a player in most versions of Indigenous policy futures that we might imagine. In their chapter on 'Treaty as a Pathway to Indigenous Controlled Policy: Making Space, Partnering, and Honouring New Relationships', Sarah Maddison and Anya Thomas explore the process and practice of treaty making, and the dilemma that formal negotiations will pose for settler colonial societies. Their work begins by describing a profound shift in Indigenous-settler relations, as four Australian jurisdictions—Victoria, Queensland, Tasmania, and the Northern Territory—have recently committed to treaty processes with First Nations. While these are still nascent processes, there is a growing realisation of the profound impact that treaty will have on public policy-making. State bureaucracies that have understood their role as primarily one of service delivery to First Nations will have to reorient themselves to become treaty partners with First Nations seeking to exercise greater control and autonomy. Foundationally, Maddison and Thomas note that the practices of settler state control are incommensurable with the rights and responsibilities of Indigenous social and political life; genuine treaty, they argue, cannot be about one sovereign entity defining the other but rather requires a reconceptualisation of how 'sovereignty' itself is defined. If sovereignty is understood as a property of people rather than of institutions, then modern treaty-making and the public policy approaches that can subsequently emerge do contain the potential to be enabling for First Nations' futures. Maddison and Thomas explore policy co-design, approaches to resurgence and refusal, and the centrality of land to identity, and in doing so explore opportunities and challenges to potential future relations between First Nations and the state, adopting an incremental approach to the development of new social and political norms. Their chapter explores the possibilities and challenges of transforming public policy-making through treaty, arguing that although it will take time to recast the terms of engagement, a genuinely Indigenous-controlled policy can become the new political norm.

In the final chapter in this volume Miriam Jorgensen, Alison Vivian, Anthea Compton, Donna Murray, Debra Evans, and Janine Gertz turn our attention to the urgency of this moment in time. In their chapter 'Yes, the Time is Now: Indigenous Nation Policy Making for Self-Determined Futures' they write about the history of the term 'Indigenous Nation Building' (INB) and the utility of the approach for three Aboriginal nations in Australia; The Gunditjmara People (Victoria), the Gugu Badhun Nation (Queensland) and the Wiradjuri Nation (New South Wales). The authors demonstrate how the logic of settler colonialism in public policy translates to a focus on the actions of the state, and thus to a singular focus on transforming those actions, as the only measure of success in Indigenous affairs. This obscures how Indigenous peoples are agents and actors, are rights-holders, enacting their self-determination through their own 'Indigenous policies' in an adverse sociopolitical context. Like Graham and Brigg, Jorgensen et al. are concerned with issues of authority, strategy, and temporality; the promise of INB lies in its approach to strengthening the collective identities of Indigenous peoples and securing Indigenous futures for those identities and ways of being that do not rely on the state for permission to thrive.

Together, the chapters in this volume point us towards a future that may yet break with the chaos and control of the settler colonial order. This future is far from assured, but a pathway is visible in these case studies of resilience, resistance, refusal, and resurgence—and of course in myriad other cases not captured here. That this pathway exists at all is testament to First Nations' determination and creativity in the face of settler colonial logics of control. Indigenous peoples work with and against the state, protecting the old and imagining the new, on their own terms, in their own institutions, through their own values and priorities, to make policy that is both by and for Indigenous peoples.

## References


Aitken, W. A. (2009). Indigenous policy failure and its historical foundations. *International Journal of Critical Indigenous Studies*, *2*(1),15–23.

Alfred, T., & Corntassel, J. (2005). Being indigenous: Resurgences against contemporary colonialism. *Government and Opposition*, *40*(4), 597–614. https://doi.org/10.1111/j.1477-7053.2005.00166.x

Anderson, I. (2015). The Crisis of Australia's Indigenous Policy. *Meanjin*, *74*(3), 54–59. https://doi.org/10.3316/informit.620556605198824

Anderson, P., & Leibler, M. (2017). *Final report of the Referendum Council*. Commonwealth of Australia.

Baldwin, A. (2012). Whiteness and futurity: Towards a research agenda. *Progress in Human Geography*, *36*(2), 172–187. https://doi.org/10.1177/0309132511414603

Begay, M. A., Cornell, S., Jorgensen, M., & Kalt, J. P. (2007). Development, government, culture: What are they and what do they have to do with rebuilding native nations? In M. Jorgensen (Ed.), *Rebuilding native nations: Strategies for governance and development* (pp. 34–56). University of Arizona Press.

Carlson, B., & Frazer, R. (2021). *Activism*. Palgrave Macmillan. https://doi.org/10.1007/978-3-030-84796-8_8

Commonwealth of Australia. (2021). Indigenous Voice Co-design Process Final Report to the Australian Government, National Indigenous Australians Agency, Canberra. https://voice.niaa.gov.au/sites/default/files/2021-12/indigenous-voice-co-design-process-final-report_1.pdf

Cornell, S. (2015). Processes of native nationhood: The indigenous politics of self-government. *International Indigenous Policy Journal*, *6*(4), 1–27. https://doi.org/10.18584/iipj.2015.6.4.4

Cornell, S., & Jorgensen, M. (2019). What are the limits of social inclusion? Indigenous peoples and indigenous governance in Canada and the United States. *American Review of Canadian Studies*, *49*(2), 283–300. https://doi.org/10.1080/02722011.2019.1613790

Cornell, S., & Jorgensen, M. (2020). Indigenous nations in postracial America: Rethinking social inclusion. *The Review of Black Political Economy*. https://doi.org/10.1177/0034644620966033

Jorgensen, M. (Ed.). (2007) *Rebuilding native nations: Strategies for governance and development*. University of Arizona Press.

Grande, S., & McCarty, T. L. (2018). Indigenous elsewheres: Refusal and re-membering in education research, policy, and praxis. *International Journal of Qualitative Studies in Education*, *31*(3), 165–167. https://doi.org/10.1080/09518398.2017.1401144

Laws, J. (Presenter), & Howard, J. (Guest). (2000, May 19). *Interview with John Howard, Prime Minister of Australia*. [Radio broadcast]. 2UE Radio. https://pmtranscripts.pmc.gov.au/release/transcript-22788

Maddison, S. (2019). *The Colonial Fantasy: White White Australia Can't Solve Black Problems*. Allen and Unwin.

Maddison, S., & Brigg, M. (Eds.). (2011). *Unsettling the settler state: Creativity and resistance in Indigenous settler-state governance*. Federation Press.

Patterson, M. (2017). *Commonwealth Machinery of Government in Aboriginal and Torres Strait Islander Affairs: 50 years of Commonwealth Public Administration in Aboriginal and Torres Strait Islander Affairs*. IAG Working Paper Series. No. 1, Aboriginal and Torres Strait Islander Affairs Group, Canberra: Department of the Prime Minister and Cabinet

Roche, G., Maruyama, H., & Virdi Kroik, Å. (2018). Indigenous efflorescence: Beyond revitalisation in Sapmi and Ainu Mosir. ANU Press.

Simpson, L. B. (2016). Indigenous resurgence and co-resistance. *Critical Ethnic Studies*, *2*(2), 19–34. https://doi.org/10.5749/jcritethnstud.2.2.0019

Simpson, A. (2017a). The ruse of consent and the anatomy of 'refusal': Cases from indigenous North America and Australia. *Postcolonial Studies*, *20*(1), 18–33. https://doi.org/10.1080/13688790.2017.1334283

Simpson, L. B. (2017b). *As we have always done: Indigenous freedom through radical resistance*. University of Minnesota Press.

Smith, D., Wighton, A., Cornell, S., & Delaney, A. V. (Eds.). (2021). *Developing governance and governing development: International case studies of indigenous futures*. Rowman and Littlefield.

Strakosch, E. (2019). The technical is political: Settler colonialism and the Australian Indigenous policy system. *Australian Journal of Political Science, 54*(1), 114–130. https://doi.org/10.1080/10361146.2018.1555230

Tuck, E., & Gaztambide-Fernández, R. A. (2013). Curriculum, Replacement, and Settler Futurity. *Journal of Curriculum Theorizing, 29*(1), 72–89. https://journal.jctonline.org/index.php/jct/article/view/411

Turnbull, M., Brandis, G., & Scullion, N. (2017, October 26). *Joint media release: Response to Referendum Council's report on Constitutional Recognition* [Press release]. https://parlinfo.aph.gov.au/parlInfo/search/display/display.w3p;query=Id%3A%22media%2Fpressrel%2F5596294%22

Vizenor, G. (2008). Aesthetics of survivance: Literary theory and practice. In G. Vizenor (Ed.), *Survivance: Narratives of Native presence* (pp. 1–23). University of Nebraska Press.

Wolfe, P. (2006). Settler colonialism and the elimination of the native. *Journal of Genocide Research, 8*(4), 387–409. https://doi.org/10.1080/14623520601056240

# Chapter 2
# Indigenous Public Policy Futures: A Manifesto for Relationalist Public Administration

**Mary Graham and Morgan Brigg**

**Abstract** By taking a long view of public administration on the Australian continent, this chapter places the recently imposed British-European political system alongside ancient Indigenous governance. We argue that imposed settler administration struggles to gain foundational legitimacy in ways that require turning to a relational orientation, as conceptualised by Aboriginal Australian peoples, to support Australian nation building and the accompanying provision of public goods. This relational orientation supports a threefold agenda: the development of a viable political settlement to address the colonial wound that haunts the Australian polity; supporting the development of effective forms of public administration in relationship with this land, and; responding to questions of identity and belonging by developing forms of governance that begin by engaging with first peoples' political ordering and forms of governance of this place rather than with their recently arrived and relatively ungrounded European counterparts. We argue that this relational orientation, borne of original 'civilizational culture', can inform the provision of public services by Australian governments and hence the development of an Australian 'civilisational state'.

**Keywords** Relationality · Aboriginal and Torres Strait Islander · Civilisation · Public administration · Indigenous governance · Australia · Settler colonialism

## 2.1 Introduction

For tens of thousands of years prior to European arrival, Australia's first peoples developed and practiced sophisticated and grounded political systems for governing human relations on the Australian continent. More recently, a British-European political system has asserted political jurisdiction through force and by dispossessing Indigenous people. The ensuing forms of public administration derive almost entirely from imported British-European liberal ideas and prescriptions that struggle

M. Graham · M. Brigg (✉)
The University of Queensland, St Lucia, QLD, Australia
e-mail: m.brigg@uq.edu.ua



N. Moodie and S. Maddison (eds.), *Public Policy and Indigenous Futures*, Indigenous-Settler Relations in Australia and the World 4,
https://doi.org/10.1007/978-981-19-9319-0_2

to engage with Indigenous people and political ordering other than through settler colonial logics. This dynamic appears entrenched and yet settler political order and forms of governance are invariably marred by the illegitimacy that comes with their means of acquisition. Settler administration cannot mount a compelling answer to the questions 'by what lawful authority do you come to our lands? What authorises your efforts to dispossess us?' (Watson, 2012, p. 12). Settler sovereignty exists de facto, but into the future and in the face of evolving expectations of legitimate governance, it will struggle to exist *de jure*. Nor can Settler governance and administration call upon either a longstanding civilizational tradition (such as the French or Chinese) or other form of foundational legitimacy. In its current form, settler administration cannot *take hold* in this place; it remains an insecure and uncertain mechanism governing an ungrounded European outpost.

This chapter takes a long view of public policy and administration on the Australian continent. Rather than accept the assertion of Settler jurisdiction, we begin with the priority of Aboriginal and Torres Strait Islander peoples as owners and runners of 'Country'. Irene Watson's question, 'by what legal authority…', hums in the background as we argue for the value of 'relational' public administration. We argue for relational public administration not only in Indigenous affairs policy (and thus as a venture circumscribed by the rubric of self-determination) but for the continent as a whole and thus for all peoples who seek to make the continent home. Our argument resonates, by extension rather than a claim to universality, to other similar jurisdictions. Our case for relational public administration is not built on the fact of the precedence of Indigenous peoples, or on a normative argument that attending to the governance schemas of first peoples is just and right, though these may readily lend support to our case. Rather, our argument is that a relational orientation, as conceptualised by Aboriginal Australian peoples, supports the provision of public goods that are necessary for Australian nation building.

We suggest that a relational orientation is necessary to support a threefold macro agenda. First, it is necessary to underpin the development of a viable political settlement to address the colonial wound that haunts the Australian polity by supporting an authentic relationship between original and newcomer peoples. Second, a relational orientation can underpin the development of effective forms of public administration in relationship with *this land* that all life on the Australian continent rely upon for sustenance and wellbeing. The ensuing benefits range from the technical and practical, including matters of appropriate resource use and social inclusion and wellbeing, to an ability to grapple more effectively with ineffable questions of identity and belonging. Finally and overall, if public administration on this continent is to belong, it is necessary to develop Australian forms of governance that begin by engaging with first peoples political ordering and forms of governance *of this place* rather than with their recently arrived and relatively ungrounded European counterparts.

Our short chapter cannot fill out all that are necessary for pursuing the foregoing agenda. That work is being undertaken, often in underappreciated and subterranean ways, in other writing, in practices and in relationships. Here we make some offerings to this wider effort in three parts. First, we introduce selected principles—both abstract and grounded—to indicate how important Indigenous philosophical precepts might resonate in the present. This has us discuss relational balance and accompany ethical principles, including gender balance. Second, we consider how such principles can manifest in 'modern'/introduced governance systems by considering the case of the National Congress of Australia's First Peoples and then how the notion of 'Caring for Country' resonates more widely. Finally, we return to broader matters of how Aboriginal 'civilizational culture' might inform the public services of Australian governments, and thus the development of an Australian 'civilisational state'.

## 2.2 Relational Balance

Aboriginal peoples achieved long-term order and security on the Australian continent through the development of systems of co-existence with the natural environment and other-than-human beings. The conceptual architecture of this system is complicated and multifaceted, but one succinct way of indicating its core character is to highlight the foregrounding of 'relationalism' contra 'survivalism'—and especially the mobilisation of relationalism to deal with the all-too-human difficulties and challenges of survivalism (Brigg et al., 2021). All peoples necessarily grapple with 'survivalist' tendencies—with egoism and the impulse to competition that emerge alongside the tensions and demands of human interaction. Aboriginal peoples developed institutions to manage these challenges without resorting to the domination and hierarchy of sovereignty that is familiar to European conceptualisations and practices of political ordering and public administration. Relationalism is not the opposite of survivalism, and relationalism does not seek to either subdue or banish survivalist tendencies. Rather, relationalism attends to and manages the problems that survivalism generates while recognising it as part of being human.

Relations are laterally networked in Aboriginal political ordering, but a primary relation is between people and land or sea (with this conjunction termed 'Country' in Aboriginal English). Other relations, including those with other-than-human beings, are linked with the relationship between people and land and sea. People are ultimately, inextricably, and forever obliged to sacralised land and/or sea for existence. Amidst the ensuing relations that generate and secure individual and group wellbeing and security, obligation develops through reciprocal arrangements in which human collectives—families, clan and language groups—'look after' land and/or sea within local regions. These relations are ordered through a complex systematisation—the Law or 'Dreaming'—embedded within and among individuals and groups, providing a poetic and self-organising regulatory system for guiding behaviours and relations among people (Stanner, 1979, pp. 23–28). The Law orders the world in myriad ways, including through cultural narratives, kinship systems, and ceremony that develops spiritual integrity, and recognises land and/or sea as a moral entity by enacting duties to ancestors. More prosaically, the Law provides injunctions for managing behaviour and being a lawful being as well as guiding the primacy of responsibilities to family, ancestors, and Country, including through totemism.

A core part of lawful relational behaviour is the pursuit of continuity (and thus long-term order and security) through balance. Deborah Bird Rose explains how balancing relations through totemism manages common property in many parts of Aboriginal Australia. The cross-cutting dreaming trails of ancestors sacralise land, restricting and guiding human behaviour in ways that provide 'structures of restraint, management for long-term productivity, control of sanctuaries, protection of permanent waters, refugia, breeding sites, and [protection] of certain plant communities' (Rose, 2013, p. 138). At the macro-level, totemism establishes a balance between people and land. However, the achievement and operationalisation of this balance is also pursued through complementarity and balance in human and kinship relations. Although rights to forage are widespread, for instance, the responsibilities to land and ancestors that come 'with that right differ depending on one's relationship to the country' (2013, p. 138). These responsibilities are in turn organised complementarily within the kinship system. One group—for example, relations on one's mother's or one's father's side—may be the 'owner' of a particular practice (a ceremony or country maintenance such as burning) while the other complementary group is the designated 'manager' of the practice.

As Rose highlights, these arrangements for resource management are 'part of a system for organising of difference in the service of producing interdependence'—difference 'is organised to be complementary rather than oppositional, and thus is constitutive of cultural, social and ecological life' (Rose, 2013, p. 138). These types of complementary interdependence arrangements for providing relational balance resonate and concatenate throughout Aboriginal socio-political ordering to provide for long term and secure ordering. Crucially, relational balance does not imply an absence of conflict or tension—this is not the peaceful and 'close to nature' utopia that is sometimes imagined in primitivistic renderings of Indigenous peoples. Maintaining balance can be conflictual frustrating and agonistic, and this can be testing, trying and time-consuming.

Gender relations are another realm in which difference is organised to produce interdependence and balance. While early observers of Aboriginal and Torres Strait Islander communities witnessed the influence of elder men, it is likely that these observations were themselves at least partly unbalanced, reflecting the gender hierarchies of the incoming colonial society. Most early European observers of Aboriginal society were men, and it now seems likely that they mistook their predominant contact with men (reflecting the often-separate domains of gendered life in Aboriginal groups) alongside their own interests in matters of the 'public realm' as code for the dominance of older men, sometimes termed a gerontocracy. However, women certainly have access to Law and Dreaming in Aboriginal Australia, with men and women having separate yet overlapping jurisdictions and responsibilities: 'men and women have distinctive roles to play but each has access to certain checks and balances which ensure that neither sex can enjoy unrivalled supremacy over the other' (Bell, 2002, p. 182).

Aboriginal men—as well as women—hold this understanding of relational balance. Diane Bell notes that an Aboriginal man she dealt with was 'concern[ed] to balance the correct consultation procedures with respected female leaders against his own self image as a "leader" in the eyes of white male advisers or politicians' (2002, p. 29). Even more pointedly vis-à-vis newcomer observers of Aboriginal gender relations Bell recounts that:

> several older men approached me, rather hesitantly, and told me that they too had some ritual business which I might see if I wished. They quickly added that they knew the women were teaching me properly but I might like to know that they were also important in the upholding of the Law. Women weren't the only ones with Law, they told me, thereby standing the concept of male dominance on its head. (Bell, 2002, p. 34)

This is not an argument that no gendered power relations are at play, but instead an indication that there is substantial scope for recasting simplistic introduced understandings of gender relations among Aboriginal people.

Relational balance only explains a little about the sources of Aboriginal political order and security on the Australian continent, but it is one way of parsing a key principle of Aboriginal Australian socio-political ordering. The operation of relationalism and other principles of Aboriginal political ordering have been disrupted and disavowed, pushed aside and underground by settler colonialism, but continue nonetheless. It is thus unsurprising that relational balance re-emerges in efforts to represent Aboriginal peoples in conditions of political asymmetry in ways that indicate possibilities for drawing on Aboriginal principles more broadly.

## 2.3 (Re-)Applying Aboriginal Principles

Settler colonial incursion has violently interrupted tens of thousands of years of Aboriginal socio-political ordering on the Australian continent. While early encounters on the frontiers did sometimes involve mutual accommodations, these were replaced by highly unequal relations institutionalised by colonial lawmakers and public servants in public administration. These relations required Aboriginal people to trade in the political ontology and currency of the introduced political order. As dominant relations began to be more widely challenged with late-1960s social movements, there arose, *inter alia*, a succession of aspirational Aboriginal consultative and representative bodies, from the National Aboriginal Consultative Committee (NACC) to the Aboriginal and Torres Strait Islander Commission (ATSIC), which was disbanded by the Australian Government in 2004–2005. Despite social movement advocacy and the shift to a policy of 'self-determination', none of these bodies were Aboriginal-controlled. Nor did any of the bodies involve 'explicit recognition of any separate source of political authority' aside from that of the introduced state (Bradfield, 2006, p. 84), though Sanders argues that with ATSIC, 'it seemed possible to think of law and governmental authority in Australia as flowing from Indigenous sources' (Sanders, 2018, p. 114).

Here we cannot fully consider how to address the complicated and entangled polity-to-polity relations that impinge upon reworking Aboriginal principles for political ordering to move beyond the current messy interregnum toward an era of legitimate political administration on the Australian continent, and nor can we resolve questions about Indigenous representation. Instead, we want to draw attention to what can and does happen, amidst political asymmetry, when Aboriginal and Torres Strait Islander people innovate, even with a hostile setting. In particular, we discuss how principles of relational balance were brought to bear in the development of an Indigenous-led representative body from 2009, and hence how Aboriginal principles can be mobilised in contemporary organisations and public administration.

In the wake of the demise of ATSIC, and given creeping and then forceful recognition of the policy need for an Indigenous representative body, in the Australian Government engaged the office of the Social Justice Commissioner to convene national consultations to provide advice on a model for a new national representative body (Calma & Dick, 2011, pp. 170–172). The resulting report proposed a model for a new national representative body independent of government, though funded by government in the first instance (Australian Human Rights Commission, 2009, p. §3). This aspiration was borne out, and the new body, the National Congress of Australia's First Peoples, came into being in 2010 as a public company rather than a statutory authority.

Our claim is not that the consultation process or the resulting structure wholly reflect and manifest relational balance, and certainly not that either are perfect: both the consultation and the structure have been subject to critical questioning (see Muir, 2010). The consultations also faced crucial decisions such as 'whether the role of the National Representative Body would be to represent a national perspective on Aboriginal and Torres Strait Islander issues or to represent Aboriginal and Torres Strait Islander peoples across the nation'. Such invidious decisions are destined to result in imperfections (Calma & Dick, 2011, p. 175). Rather, we claim that important elements of the structure of the Congress reflect an attempt to manifest relational balance in the context of an extant highly asymmetrical political ecology largely determined by the dominant settler system. Of course, it is hard to disentangle attempts to assert Indigenous-specific forms of governance from efforts to develop a more effective and secure institutional life in the interface with—and hence on the terms of—the settler state. We nonetheless posit that this attempt reflects genuine Indigenous innovation that does not derive wholly from the settler order and that might be valuably brought to bear more widely. So while the Congress went into voluntary administration in 2019 after being defunded by the Australian Government, elements of its structuring evoke the possibilities of drawing upon Aboriginal principles in contemporary governance.

The Congress consisted of three chambers and a powerful ethics council (which one of us, Mary, was a member of from 2011—2017) alongside a small national executive. One chamber drew delegates from peak representative bodies, the second drew delegates from 'Sectoral Peak Bodies and experts', and the third chamber drew delegates from among Aboriginal and Torres Strait Islander individuals and organisations not otherwise qualifying in either of the Chambers of the organisation (Australian Human Rights Commission, 2009, p. 24). There were limits on the number of delegates from any single organisation and all delegates were vetted by the ethics council that took a type of 'elders council' role providing independent advice on standards. The decision to use a delegates model accompanied by an ethics council reflects a lack of faith in electoral processes to deliver good governance, reflecting Coombs' observation that this model 'seems closer to traditional Aboriginal processes than those based upon the models of representative parliamentary institutions' (Coombs, 1994, p. 132).

This combination of three chambers and an ethics council of reputable individuals represents an attempt to effect interdependence and relational balance among Indigenous interests within the extant political ecology and in the context of the aim to provide a national perspective. Prospective members of the Congress were nominated 'from below' from the Indigenous community and sector (primarily through organisations), but then vetted by the ethics council. They then came together to from the Congress that then selected the national executive (with nominees also vetted by the ethics council).

Perhaps the most striking evidence of commitment to relational balance in the Congress structure rests in the commitment to gender balance or equality (Muir, 2010, p. 1991). This balance is reflected throughout the organisation—in each of the chambers, on the ethics council, on the national executive, and in the form of co-chairs of the Congress. Such a thoroughgoing commitment to gender balance is unprecedented in any cognate settler institution or governance arrangement, leading to the conclusion that this is at least partially an Indigenous innovation within the Australian setting.[1] Moreover, having no single figurehead symbolises the commitment to relational balance (and accompanying processes and modes of decision-making) that designers of the organisation intuitively or explicitly sought to produce.

In some respects, these innovations are modest. Indeed, we are not asserting that the Congress represents a 'model', or even that the specific types of balancing that we have highlighted in the Congress structure should be given precedence. However, considered in the context of previous bodies that created a bureaucratic representative structure determined by government and the constraints of the extant political ecology, the bottom-up representation and balancing among the key elements of Congress deserves to be viewed as a significant innovation. With this we are signalling, first, that even in highly asymmetric and difficult political circumstances, relational Aboriginal political commitments can and do resurface. In the metaphor of palimpsest, 'older images and meaning lie underneath the membrane of modernity ready to re-emerge, restored and realigned' (Cowlishaw, 1999, p. 295). Second, we seek to point to how relational commitments can play out in contemporary governance arrangements in ways that contribute to overall public goods (balancing of interests and voices, and gender balance). Here the pursuit of Aboriginal political ontology does not imply the accommodation or representation of a particular sectoral interest, but rather tapping into a serious political-philosophical source with grounded legitimacy in the Australian continent.

Equally, an organisation such as Congress, constrained as it was by the extant settler-colonial political ontology, naturally exhibited aporia vis-à-vis Aboriginal

---

[1] This commitment to gender balance is also reflected in recent proposals for an Indigenous voice to government and parliament Commonwealth of Australia, N.I.A.A. (2020). Indigenous Voice Co-design Process Interim Report to the Australian Government. accessed 20/02/2021.

political-philosophical foundations. Most strikingly, the Congress' formal structures did not explicitly install Country as the primary relation between people and land or sea as a guiding principle. While this is undoubtedly an aporia, Mary's experience on the ethics council of Congress points to the way commitments to Country may nonetheless seep into governing institutions such as Congress and beyond. Mary notes that the language map of Indigenous Australia was displayed behind participants in the meetings of Congress, serving as a reminder and a type of guide about the need for Country-informed decision-making. With that as a backdrop, she was stuck by how delegates seemed to fluidly, comfortably and without fuss, (re-)turn to their old habit of running the country.

Notwithstanding the limitations of dominant settler-colonial political ecology and ongoing struggles, compromises, and accommodations, drawing upon Country to inform and drive decision-making has become increasingly widespread in (Indigenous) Australia. Many Indigenous organisations (those controlled by, based in, or primarily serving Indigenous communities) reference Country, with serving or caring for Country often explicitly referenced as the *raison d'etre* for land-focused organisations (e.g., see Quandamooka Yoolooburrabee Aboriginal Corporation, 2020, p. 16). The language of 'Caring for Country', initially introduced to public policy circles with the establishment of the Northern Land Council's Caring for Country unit in 1995 and popularised through a range government programs (AIATSIS, 2011), has become an iconic phrase for a diverse range of private, community and government ventures and partnerships (see too McCaul's chapter in this volume). Even Australian Prime Minister Scott Morrison invoked the 'principle of caring for country' to evoke, *inter alia*, 'responsible management and stewardship' in his National Press Club of 26th May 2020 (Prime Minister of Australia, 2020).

Our claim in this section is not that Aboriginal political principles are being applied without struggle, that Congress represents a perfect model, or that notions such as Caring for Country can readily transform the political landscape. Innovations such as those of Congress are, *inter alia*, always entangled with the constraints that come to bear with the assertion of defacto settler sovereignty. Our argument, rather, is that Aboriginal political principles are at play, are increasingly being recuperated, and can be drawn upon to inform contemporary governance. These phenomena signal possibilities in the long view of public administration borne of tens of thousands of years of Aboriginal governance of the Australian continent.

## 2.4 From Civilisational Culture to Civilisational State?

Contemporary Australian political relations and accompanying public administration can be characterised as a messy interregnum. Indigenous political ordering of the continent has been violently interrupted by colonialism and the European-derived assertion of sovereignty and accompanying society and political system. Yet the resulting political order, orchestrated through the settler state, cannot achieve legitimacy or belonging. The settler state exercises *power*, but in terms of Aboriginal

Law, it is an uncivilised form of *authority* (Brigg et al., 2019; Rose, 1984). The state has, *inter alia*, mobilised unjustifiable hierarchy and control in perpetuating mass harm against the original peoples of the continent. For these and a host of cognate reasons, the Australian state and accompanying public administration will continue to struggle for legitimacy until the relationship with original owners and runners of Country is appropriately addressed and governance is informed by a philosophical orientation that is appropriate to and grounded in this continent. One way to pursue this long-term work is to adopt a relational orientation to public administration, drawing upon Aboriginal civilising principles to move toward a civilisational state. Aboriginal relational commitments are necessarily modified and reshaped in this process, but this is possible, as the previous section demonstrates.

It perhaps seems old-fashioned to write of 'civilising' and 'civilisations', and yet we do so in part to upend the way that civilisational thinking continues, despite lack of acknowledgement, to underpin and structure how we think about cultural difference. The dominant and largely unchallenged European-derived story about 'civilisation' has tribalism as the earliest form of human organisation, a type of primitive unifying force that provides a foundational identity, usually through cultural or religious belief. Racial thinking supports the casting of Aboriginal Australian and other Indigenous peoples as earlier primitive versions of European collectivities, no matter that the very category of the tribe arises as effect of colonial state expansion (see Mamdani, 2012). This primitive identity politics is, in turn, a bulwark against all kinds of danger, especially that posed by other groups. Tribalism has tended to be seen as a primitive system of coercive ordering within a hierarchical structure—a cluster of family groups within a cluster of clans that make up the formation of the tribe with a leader, chief or paramount. The system is seen as primitive, in part because the individual was seen as having less importance than the tribe/clan/group; he/she was seen as not free. In the dominant story, reason overcomes this type of primitive identification, subordinating it through secularised sovereignty, and accommodating it through multicultural politics of recognition.

The use of this dominant narrative to characterise the original peoples of Australia as 'tribal' seems strange from an Aboriginal Australian viewpoint, and this strangeness is borne out in the scholarly anthropological literature (see Rumsey, 1993), especially for how Western assumptions about the relationships between land, language and people tend to not apply. In particular, the primacy of connection to land and sea does not establish an exclusive tribal-territorial relation vis-à-vis other groups. Trade, out-marriage and regional relations were commonplace prior to European arrival and now abound across the continent. And commitments to relational balance, explicated above, do not translate to tribal obedience or conformity, with Aboriginal commitments to personal autonomy arguably stronger than those of European-derived individualism. From an Aboriginal perspective, the designation 'tribal' to refer to a relatively unified group organised in a hierarchical political structure seems rather more apposite for describing key features of the European nation-state, especially the state behaviours sanctioned and excused by mainstream 'Realist'

International Relations scholarship. The nation-state, especially in the early twentieth century, seems rather more tribal than any imaginable grouping of Australian Aboriginal people.

While the autonomy of Aboriginal groups is a crucial feature of Aboriginal socio-political ordering and is widely recognised in anthropological literature, this deserves to be balanced with the importance of entangled inter-group relations (also widely recognised in literature) that were discussed in the first section. Mainstream European understandings of Aboriginal socio-political ordering often oscillate between inappropriate homogenisation of Aboriginal groups to a false recognition of discrete diverse groups (often as part of efforts to counter inappropriate homogenisation). Such oscillation is likely borne of commonplace categorical and either/or thinking that wants to designate people as 'one or the other'. A more useful approach is to recognise that Aboriginal groups are *simultaneously autonomous and relationally resonant* such that there are *systemic* inter-polity relations among autonomous groups that constitute groups as mutually implicated. These relations are continent-wide; a heaving body of interconnected polities as represented in David Mowaljarlai's map (1993).

In this schema neither 'tribe/s' nor 'state' will suffice. Rather, it seems more appropriate (and perhaps useful) to refer to Aboriginal peoples' sophisticated political ordering of governance relations over tens of thousands of years through relational balance (and accompanying forms of patterning) as civilisational, or as a 'civilisational culture'. In this context, it is the 'state' that deserves to be seen a foreign form of political organisation. Moreover, it appears as a particularly unruly, 'wild', and immoral form of politics that needs to be civilised to the Australian continent (Brigg et al., 2019; Rose, 1984). We are suggesting, then, that the encounter between a civilisational culture and a tribalist immoral state is fertile ground for a mutually beneficial development of a 'civilisational state' as part of a rethought process of nation building.

We derive the notion of civilisation*al* state from the term 'civilisation state' used in recent decades by countries including China, India, Turkey, and Russia to stress their civilisational identity in contrast to more narrowly grounded twentieth century 'nation-state'. We adopt the suffix -al to sidestep what Amitav Acharya calls regime-based attempts to mobilise civilisation-state ideology in contrast to liberal Western values and to counter Western dominance (Acharya, 2020). Rather than trading in identity politics, we suggest that the relational values embedded in Aboriginal political ordering can find traction with newcomers to the Australian continent and thus could become shared foundations for legitimate governing by all who live here. There are no short-term prospects of turning the settler state into a thoroughgoing relational governing apparatus, but the státe does exhibit some relational governance phenomena and in a long view of public administration, the civilising influence of Aboriginal political culture can be brought to bear to develop a uniquely Australian civilisational state.

By way of example, one avenue for developing long-term shared foundations could lie in the development and articulation of ideas of 'stewardship'. The ethical relation underpinning stewardship has individuals in relationship with and responsible for others and phenomena larger than their selves. This commitment is central to an Aboriginal custodial ethic that is central to Caring for Country. Meanwhile, stewardship is a public service principle that is 'currently scattered throughout the *Public Service Act 1999*' (Commonwealth of Australia, 2019, p. 22) But if the principle of stewardship were to be 'deeply embedded across the service' rather than scattered within governing documents, it needs development to identify 'why the APS exists, what it seeks to achieve, and for whom' (Commonwealth of Australia, 2019, pp. 22, 307). Addressing these questions requires discussions about who people (individuals and all Australians) are and where they live. These matters cannot be resolved without a serious engagement with original peoples and their ideas of stewardship.

The principle of relational balance that we introduced above is thus an essential ingredient, a kind of vital force and architectural principle for both running the country in Aboriginal terms and for governing that would seek to develop a civilisational state on the Australian continent. To represent this thinking is challenging at the current juncture. Aboriginal political theory as scholarly pursuit is in its early stages. Prior to colonisation Aboriginal peoples had no need to describe themselves in such terms, and until recently settler colonialism has (on the whole) disavowed any serious recognition of Aboriginal political thought. As this work gets underway (e.g., see Brigg & Graham, 2020a, 2020b), mainstream settler colonial thought is challenged to engage with a relational 'system that has no centre' (Rose, 2000, p. 220).

Amidst this complexity arises a series of conceptual tools such as the capacity for weighing and scaling relationships in balance. The pursuit of 'proportionality' (see Brigg and Graham, 2020a, 2020b) assists a society and culture to see and understand how all life forms, including humans, fit with each other. This type of fractal relationality examines how one thing fits with another from the very small to the cosmological, from the internal and the external, and from the prosaic to the ethereal with a view to seeking and establishing what security, order and meaning might be. The relational precepts that inform proportionality and a range of other Aboriginal political concepts are enduring and well established, but not rigid or unchanging. They can be put to use, as we showed in the previous section, amidst public administration and related deliberation about political ordering.

Aboriginal civilisational culture introduces not only the concept of relational balance but also the language and ethos of relationalism. This ethos invites people away from the risk of tribalism—including the political tribalism of contemporary times. Because Aboriginal relational balance and associated political concepts are enduring and confidently grounded in the continent, they can be drawn upon to invite interruption or disruption of tribalism, identity politics, racism, and ideology as well as de facto settler sovereignty. In *this* civilising schema, differences and opposites, whether between individuals or groups or concepts, are not set against each other. Instead, Country brings them into interdependent relational balance. In the pursuit of a civilisational state, public administration can be a key vehicle for advancing—and restoring—relational balance.

## 2.5 Conclusion

By taking long view of public policy and administration on the Australian continent, this chapter has paid attention to Aboriginal political ordering to advance a manifesto for relational public administration. Proceeding in this way is necessary to address the foundational question about the illegitimacy of the Australian polity deriving from its means of acquisition (Watson, 2012). Yet doing so also provides ways to advance a viable political settlement to address the colonial wound that haunts the Australian polity, to support to effective provision of public goods, and to develop a form of public administration that belongs by being grounded in *this place.* To introduce and illustrate the value of an Aboriginal Australian relational approach we briefly explicated relational balance, including gender balance, before considering how these principles manifested in the National Congress of Australia's First Peoples and the wider circulation of Country. In the final section, we turned to the possibility of how Aboriginal 'civilisational culture' might inform the development of an Australian 'civilisational state'.

Relational balance speaks to the embedding of people with a sentient landscape, with each other, and other-than-human beings to secure effective personal and community security. It is intimately concerned with questions of public good, governance and administration. Because Australian ideas of formal public administration have come from elsewhere and are relatively young, they are not secured in the context of a civilisational culture. Settler Australians have no grounded ancient knowing to call upon to underpin thoroughly the values of public service and practice of public administration. Yet the political concepts of Aboriginal civilisational culture are also amongst us through the intimate entanglement of Aboriginal and newcomer peoples (European descendants and migrants). This is especially so in myriad Indigenous-Settler relationships, personal and professional, in families and in organisations, but it is also the case in public administration.

For some the notion of pursuing relational public administration may come up against perceived irreconcilable differences between Aboriginal and settler colonial political administrative values or worldviews. This is an understandable view, but as we have argued, it is not consistent with Aboriginal relational principles that seek to balance opposition and tensions. While it is true that significant differences exist, these can be worked with as part of the pursuit of relational balance and long-term security. Aboriginal principles and commitments cannot be simply intuited, assumed, or appropriated by newcomers, but Aboriginal relationalism can sit alongside other public administrative commitments to notions such as merit and stewardship.

The Australian continent is currently occupied by a settler governmental regime that cannot secure its own civilisational foundations, including a legitimate, durable, and sustainable public administration. The pursuit of relational public administration provides a way of addressing these fundamental issues while constructing an authentic relationship between Aboriginal and newcomer peoples and developing a truly Australian form of public administration.

## References


Acharya, A. (2020). The myth of the "civilization state": Rising powers and the cultural challenge to world order. *Ethics and International Affairs, 34*(2), 139–156.

AIATSIS. (2011). *The benefits associated with caring for country: Literature review*. Australian Institute for Aboriginal and Torres Strait Islander Studies. https://aiatsis.gov.au/sites/default/files/research_pub/benefits-cfc_0_2.pdf

Australian Human Rights Commission. (2009). *"Our future in our hands"—Creating a sustainable national representative body for aboriginal and Torres Strait Islander peoples* [Report of the Steering Committee for the creation of a new National Representative Body]. Australian Human Rights Commission.

Bell, D. (2002). *Daughters of the dreaming*. Spinifex Press.

Bradfield, S. (2006). Separatism or status-quo?: Indigenous affairs from the birth of land rights to the death of ATSIC. *Australian Journal of Politics and History, 52*(1), 80–97.

Brigg, M., & Graham, M. (2020a). The relevance of aboriginal political concepts: Country, place, and territory. *ABC Religion and Ethics*.

Brigg, M., & Graham, M. (2020b). The relevance of aboriginal political concepts: How "proportionality" can help close the gap. In: *ABC religion and ethics*. Australian Broadcasting Corporation.

Brigg, M., Graham, M., & Murphy, L. (2019). Toward the dialogical study of politics: Hunting at the fringes of Australian political science. *Australian Journal of Political Science, 54*(3), 423–437.

Brigg, M., Graham, M., & Weber, M. (2021). Relational indigenous systems: Aboriginal Australian political ordering and reconfiguring IR. *Review of International Studies*, 1–19.

Calma, T., & Dick, D. (2011). The National Congress of Australia's first peoples: Changing the relationship between aboriginal and Torres Strait Islander Peoples and the State? In: S. Maddison & M. Brigg (Eds.), *Unsettling the settler state: Creativity and resistance in indigenous settler-state governance* (pp. 168–188). The Federation Press.

Commonwealth of Australia. (2019). *Our public service, our future: Independent review of the Australian Public Service*. Department of the Prime Minister and Cabinet. https://www.pmc.gov.au/sites/default/files/publications/independent-review-aps.pdf

Commonwealth of Australia, N.I.A.A. (2020). Indigenous voice co-design process interim report to the australian government.

Coombs, H. C. (1994). *Aboriginal autonomy: Issues and strategies*. Cambridge University Press.

Cowlishaw, G. (1999). *Rednecks, eggheads and blackfellas: A study of racial power and intimacy in Australia*. Allen and Unwin.

Mamdani, M. (2012). *Define and rule: Native as political identity*. Harvard University Press.

Mowaljarlai, D., & Malnic, J. (1993). *Yorro yorro: Everything standing up alive*. Magabala.

Muir, S. (2010). The new representative body for aboriginal and torres strait islander people: Just one step. *Australian Indigenous Law Review, 14*(1), 86–99.

Prime Minister of Australia. (2020). *Address*. National Press Club. https://www.pm.gov.au/media/address-national-press-club-260520

Quandamooka Yoolooburrabee Aboriginal Corporation. (2020). *Ngaliya Maguydan (Our Story): Annual Report 2019–2020*. http://www.qyac.net.au/docs/QYAC_Annual_Report_2020.pdf

Rose, D. B. (1984). Saga of captain cook: Morality in aboriginal and European law. *Australian Aboriginal Studies, 2*, 24–39.

Rose, D. B. (2000). *Dingo makes us human: Life and land in an Aboriginal Australian culture*. Cambridge University Press.

Rose, D. B. (2013). Common property regimes in Aboriginal Australia: Totemism revisited. In: P. Lamour (Ed.), *The governance of common property in the Pacific Region* (pp. 127–143). ANU Press.

Rumsey, A. (1993). Language and territoriality in aboriginal Australia. In M. Walsh & C. Yallop (Eds.), *Langauge and culture in aboriginal Australia* (pp. 191–206). Press.

Sanders, W. (2018 ). Missing ATSIC: Australia's need for a strong Indigenous representative body. In: D. Howard-Wagner, M. Bargh, & I. AltamiranoJiménez (Eds.), *The neoliberal state, recognition and indigenous rights: New paternalism to new imaginings* (pp. 113–130). ANU Press

Stanner, W. E. H. (1979). *White man got no dreaming: Essays, 1938–1973*. Australian National University Press.

Watson, I. (2012). The future is our past: We once were sovereign and we still are. *Indigenous Law Bulletin, 8*(3), 12–15.

# Chapter 3
# Success and Failure in Australian Indigenous Policy: Moral Dynamics and Rhetorical Registers

**Will Sanders**

**Abstract** How are ideas of failure and success used in Australian Indigenous policy? This question came to me in 2007 when I heard a philanthropist, newly involved in Indigenous affairs, tell a simple success story. The ideas of rhetorical registers and moral dynamics helped me think about what was going on. As a more established participant and analyst of Indigenous affairs, I knew that this was a fragile success story at best, and that developments at the time threatened its continuation. But that was the last thing the philanthropist or his audience wanted to hear. They were in a rhetorical register of challenging Indigenous policy, and those of us associated with it, to do better. My reaction to this moral dynamic was to move discussion into a different, more calming rhetorical register, explaining what had been done in the past to arrive at this delicate fragile success story, as I understood it more deeply. I recognised that some other more established participants in Australian Indigenous affairs had made similar rhetorical moves in the policy debates of the previous year or three, since the Howard Government's abolition of ATSIC as a failure in 2004–2005. Northern Territory Chief Minister, Clare Martin, for example, had produced a very clear instance of the calming rhetorical register in a speech in May 2006. In this paper I will recount in more detail these events and thoughts of 2007 and develop further these ideas of rhetorical registers and moral dynamics. Policy is by nature a future-oriented, aspirational activity, trying to make our social world better. So the presence of moral dynamics and rhetorical registers within policy debates using ideas of both failure and success should be expected and welcomed. But analytic understanding of how these dynamics and registers play out over time also needs to be developed, identifying both strengths and weaknesses of different moral and rhetorical positioning.

**Keywords** Improvement · Future-orientation in public policy · Predominance of past failure arguments over success analysis

---

W. Sanders (✉)
Australian National University, Canberra, ACT, Australia
e-mail: william.sanders@anu.edu.au



N. Moodie and S. Maddison (eds.), *Public Policy and Indigenous Futures*, Indigenous-Settler Relations in Australia and the World 4,
https://doi.org/10.1007/978-981-19-9319-0_3

## 3.1 Introduction

What is Indigenous policy achieving? Is it succeeding or failing? During a four decade career as a social science academic working on Indigenous issues in Australian public policy, I have often been asked these common-sense questions. But answering them has always felt daunting, partly because of their simplicity. It all depends, I could hear myself starting to say, what you expect public policy to achieve, in what timeframes, and for whom.

Over these decades of teaching and research I have begun to formulate some other questions about success and failure in Indigenous policy that feel less daunting. How are ideas of success and failure used by participants in Indigenous policy, and to what effect? Do participants use these ideas differently to each other, or to themselves over time? These more fine-grained empirical questions start to explore what I call the moral dynamics of Indigenous policy; how it proceeds as a contentious normative field of endeavour within Australian public policy between participants with very different perspectives and modes of involvement.

As in teaching, I begin this chapter with some general theory of policy, focusing on two significant Australian writers of recent years, Hal Colebatch (2009) and Carol Bacchi (2009). Their constructivist or interpretivist accounts of policy help identify the moral dynamics of governing. I also draw briefly from Marsh and McConnell's (2010) attempt to develop a framework for establishing policy success, which allows for contrasting assessments in different policy phases and timeframes. I then present a series of accounts of failure analysis in Australian Indigenous policy since 1990, The first of these, in the Hawke years, I judge to be predictable. The second, in the Howard years, I judge at first to be convenient, but then by 2006 and 2007 rather more complex. It is through the failure analysis of these years that I begin to explore the second term in the chapter's sub-title, rhetorical registers. I identify two registers adopted by participants in Indigenous policy, the challenging and the calming. The challenging register is easier to analyse and understand, as it constantly diagnoses failure and pushes for improvement. The calming register is more complex and, to my mind, more morally interesting. Participants using this register often regard existing policy as at least partly a success, and this sometimes sits awkwardly with a moral drive for improvement. The final section of the chapter turns to the 'Closing the Gap' policy, initiated by the Rudd government in 2008 and pursued by all governments since. My late-career reaction to this policy was rather different from my early-career reaction to the Hawke government's 1987 policy of achieving statistical equality in Indigenous employment and income by the year 2000. In this more recent instance I have analysed the competing principles and alternative idioms in Indigenous policy, rather than focusing on predictable failure.

In exploring the moral dynamics and rhetorical registers of Australian Indigenous policy, this chapter does not seek to avoid hard questions about success and failure in public policy, but rather to approach them from a more dynamic and strategic perspective.

## 3.2 Policy Theory: Three Complementary Accounts, Four Forceful Themes in Tension

Hal Colebatch (2009) has identified three accounts of policy and governing, one of which is dominant and sacred. This first 'authoritative choice' account, as Colebatch calls it, is highly positivist and rationalist. Problems exist in the world, which public officials identify, and around which they collect evidence while identifying options for government action. Advised by these officials, and sometimes by external experts, holders of positions of public authority make choices about what actions government should pursue, and these choices become government policy, embodied in legislation, cabinet decisions, or authoritative government pronouncements of some lesser status (Colebatch, 2009, pp. 24–26). Elected politicians often occupy the positions of public authority making these policy choices. However, in the English language, policy is distinguished from politics as a style of government activity and decision-making. Politics is portrayed as a partisan struggle for advantage and regarded rather negatively, while policy is aspired to as a more 'detached' and rational way of making government decisions. As noted by Colebatch (2009, pp. 63–65), the ubiquitous distinction in English between policy and politics has 'a strong normative element' in which politics comes off 'second-best in the comparison', but he also argues that, despite the different 'connotations attached' to the two terms, 'it is difficult to separate them in practice'.

Colebatch notes that 'many organisations (are) involved in exercising public authority' and that 'the authoritative choice account sees them all as part of one system, exercising the will of 'the government'' (Colebatch, 2009, p. 26). This observation leads to a second account of policy and governing, which Colebatch calls 'structured interaction'. This less sacred account portrays policy as the negotiation of order and understanding among many participants at many levels of multiple organisations, and even with stakeholders outside formal organisations, like families. At the 'street-level' of public organisations interacting with clients, there is negotiation of policy order in an attempt to make 'organised activity stable and predictable' (Colebatch, 2009, p. 27). Understandings reached in these everyday bureaucratic settings are also public policy, but only occasionally are they referred to higher public authorities for their endorsement as conscious policy choices.

This second structured interaction account is often endorsed as a broader, more realistic description of governmental practice than the authoritative choice account, while not displacing the latter from its dominant and sacred status. But the structured interaction account also introduces alternative normative criteria for judging policy, such as: Who are the relevant stakeholders in this issue and have they all contributed to negotiated understandings? Colebatch suggests that in this second account policy processes are:

> likely to be fine-grained and long-running. The participants work out a resolution of one set of problems, but new problems replace them, and they mobilize their collective problem–solving skills to address the new problems. (Colebatch, 2009, p. 29)

This focus on changing problems leads to Colebatch's third account of policy and governing, called 'social construction'. The activity at the heart of this account is 'collective puzzling', about which incipient issues in society (and markets) become articulated as problems needing government attention. Colebatch notes that problems can be identified and labelled quite differently over time, in different places or by differently positioned stakeholders. Expertise can play a significant role in problem definition but can also be contested, both by alternative expertise and by generalist political philosophies/ideologies. Colebatch notes that this third account is an 'interpretive' approach, which emphasises that policy issues and problems are 'not naturally occurring, but are 'socially constructed' by the participants' as a 'discourse' with particular 'framings' (Colebatch, 2009, p. 33).

Carol Bacchi has contributed greatly to theorising and systematising this third type of account of policy and governing, with her 'what's the problem represented to be?' approach to analysing policy. She draws attention to 'an underlying assumption that policy is a good thing, that it fixes things up' (Bacchi, 2009, p. ix). This assumption entails the idea 'that something needs to be 'fixed', that there is a *problem*' (Bacchi, 2009, p. ix). The problems that policy tries to fix, however, can often be implicit. So, a first step in analysing policy is to make explicit the problems supposedly being fixed. Bacchi has advocated a 'paradigm' of 'problem-*questioning*', rather than 'problem-*solving*' (Bacchi, 2009, p. xvii).

Methodologically, she has developed six systematic questions to ask about how social issues become 'represented' as 'problems' that need to be 'fixed' by government action, and the consequences that flow from this. These are:

1. What's the 'problem' represented to be in a specific policy?
2. What presuppositions or assumptions underlie this representation of the 'problem'?
3. How has the presupposition of the 'problem' come about?
4. What is left unproblematic in this problem representation? Where are the silences? Can the 'problem' be thought about differently?
5. What effects are produced by this representation of the 'problem'?
6. How/where has this representation of the 'problem' been produced, disseminated and defended? How could it be questioned, disrupted and replaced? (Bacchi, 2009, p. xii).

Bacchi's approach has proven popular with research students, as it provides a clear generic method to follow in analysing any policy in any country, and in comparative 'cross-border' settings as well. The approach has even developed its own acronym, the 'WPR approach'; and, in governing having an acronym is a sign of becoming institutionalised (Bacchi, 2009, p. xx).

In 2010, one of Bacchi's students applied the WPR approach to the Community Development Employment Projects (CDEP) scheme, a centrepiece of Australian Indigenous policy in the employment and income support sectors from 1977 to 2009 (Gordon, 2010). CDEP began as a way of keeping large numbers of Aboriginal people in remote areas off Unemployment Benefits (Sanders, 1985). Instead, CDEP offered participants part-time employment for a local Indigenous community organisation (Sanders, 1988). From 1987, as part of the Hawke Labor government's Aboriginal Employment Development Policy, CDEP was allowed to spread to more densely populated urban and regional areas along Australia's eastern, southern and south-western seaboards. By 1993, under the Aboriginal and Torres Strait Islander Commission (ATSIC), CDEP had become a major national program offering part-time community employment to over 20,000 participants through 182 Indigenous organisations, including in Perth and Sydney (Sanders, 1993, pp. 7–9). With expansion into major urban areas, CDEP began to face accusations of racial discrimination in comparison to social security payments and full award-wage employment (Antonios, 1997; Sanders, 1997). This problematisation was addressed in 1998 by linking CDEP more clearly to the social security system. But this was, in many ways, the beginning of the end for CDEP, which at its zenith employed 35,000 participants in over 270 locations around 1997. Following an independent review under the newly-elected Howard Coalition government (Spicer, 1997), CDEP began to face more wide-ranging criticisms, including from some Indigenous leaders. As Gordon (2010) argued, CDEP was increasingly framed as part of a problem of Indigenous welfare dependency, rather than its 1970s framing of addressing a legacy in remote areas of under-award-wage employment and exclusion from the social security system. In Bacchi's terms, the problem representations and policy fixes of 1977 and 1987 were becoming reframed and represented as a different problematisations after 1997.

CDEP is not the focus of this paper, but it does play a role in all the instances of failure analysis discussed. So, introducing CDEP to illustrate policy theory around changing problem representations helps understand the relevance to success and failure analysis of this third type of account of policy and governing. Also, following Gordon, I have found Bacchi's WPR approach insightful for thinking both about CDEP's 'fall from favour' during the 2000s, and about the reframed problematisation of Indigenous employment in remote areas that has emerged since 2010 (Sanders, 2016, 2017). Problem construction, or framing, is a contingent conceptual step in public policy, that Bacchi, Colebatch, and others have powerfully drawn to our analytic attention over the last three decades, through this third type of account of policy and governing (see Fischer & Forrester, 1993 for a path-setting collection on the importance of framing).

Colebatch argues that the three accounts of policy and governing should not be seen as competing, but rather as complementing each other. They draw attention to 'different dimensions of policy practice', the 'vertical', the 'horizontal' and the 'scene-setting', which are 'not alternatives', but tend each 'to assume the others' (Colebatch, 2009, p. 35). Colebatch also talks of three themes that give the idea of policy great force: authority, order, and expertise (Colebatch, 2009, pp. 8–9). While these themes are all desirable, as well as forceful, Colebatch notes that they can be in considerable 'tension' with each other. The desire of newly elected ministers to make their own authoritative choices about government action relating to a service or industry within their portfolio may be at odds with the order already negotiated between long-standing stakeholders. These stakeholders can include both consumers and producers of particular services or industries, who operate primarily in the horizontal dimension of policy activity, as well as experts who claim deep knowledge through careers of dedication spanning decades. Different timeframes of public policy are in operation here, as well as participants with very different roles and perspectives. So, while Colebatch's three accounts and themes of policy are not in direct competition as alternative explanations of policy processes, they do in juxtaposition produce significant tensions in processes of governing.

In my research and teaching on Australian Indigenous policy, I have felt the need to add a fourth forceful theme within the idea of policy, to augment Colebatch's authority, order, and expertise. This is the idea of improvement. All public policy aims for improvement through reasoned thought and analysis. In English, as already noted, this is embedded in the linguistic distinction between policy as detached, rational, and good; and politics as more negatively regarded and partisan. A drive for improvement in government action through investigation, evidence, and reason, contributes to the sacredness of Colebatch's authoritative choice account. But the idea of improvement is often how experts assert a role in defining problems in the third type of account, and can also be invoked by deeply-involved consumer or producer stakeholders in a service or industry operating primarily through the structured interaction account. The common desire for improvement across all participants and accounts creates forceful moral dynamics in public policy. Governments want to claim to be doing better than in the past, and are always being pushed to do so. This makes public policy a very future-oriented activity, which often disowns the past. Yet institutional order established by past policy also persists. Balancing future improvement with the established order of existing practice lies behind much debate over policy success and failure. The ensuing discussion will suggest that Indigenous policy involves a particularly strong push for improvement, which is often combined with a simplistic assessment of past policy as a failure. A second order strategic question then becomes: how can success analysis be used to conserve, from the past, things worth keeping?

In a recent conceptual debate, Marsh and McConnell (2010) attempted to develop a general 'framework for establishing policy success'. They reviewed a 'policy evaluation' literature focused on 'measuring public sector improvement', then a literature that had started with 'policy fiascos' but had moved, as it matured, to examining both 'success and failure in public governance'. This latter literature noted that there could be 'programmatic' as well as 'political' questions of success and failure, and that these could be assessed very differently. Marsh and McConnell added a third 'process' dimension. However, one of the original authors in this literature suggested in reply that process vs outcome was an independent cross-cutting dimension, that could be applied to both programmatic and political assessments, creating a fourfold two-by-two space for assessing policy success and failure rather than a threefold list (Bovens, 2010).

What seems important in this debate is that there are multiple criteria for judging success and failure in public governance, and also very different timeframes over which such judgements can be made. Perhaps one of the most common patterns is to judge policy a success of political management in the short term, but to judge it more wanting in programmatic terms over longer timeframes. This is the pattern that Murray Edelman observed in the 1970s in two books entitled *Politics as Symbolic*

*Action* (1972) and *Political Language* (1977). As the sub-title of the second book suggested, there can be 'words that succeed', politically in the short term, 'and policies that fail', more programmatically in the longer term (Edelman, 1977).

## 3.3 Predictable Failure: Indigenous Employment and Incomes Policy Under the Hawke Labor Government

My early career writing on success and failure in Australian Indigenous policy identified a classic instance of words that succeed in the short term while predicting certain failure of intended programmatic outcomes over a longer timeframe. It was 1991, when, with the encouragement and assistance of Jon Altman, I critiqued the Hawke Labor government's Aboriginal Employment Development Policy (AEDP) of 1987.

The AEDP had involved strong public statements committing to statistical equality in employment and income matters between Aboriginal and other Australians by the year 2000. But analysis of the 1986 census and other sources by several social scientists suggested that this was a huge, wellnigh impossible, task (Altman, 1991). My conceptual question was: why would a government set unrealistic goals for itself, which it is destined from the outset not to meet? A partial answer could be that the government did not fully appreciate the deep structural legacies, both historical and demographic, on which its policies would be acting. But this only took the analysis so far, as there were aspects of the Hawke government's AEDP, like the expansion of CDEP, that were in fact quite realistic and conservative about what could be achieved (see Altman, 1991, pp. 11–14). A second part of the answer seemed to be that this is just what governments do; that it is a deep pattern in policy making to attempt significant improvement even if this leads to later perceptions of falling short. As I wrote back then:

> Policies ultimately may fail, but … strong policy statements actually succeed … They allow government to be perceived, for a time at least, as boldly and coherently tackling important social problems, even if the cost of this short-term symbolic success is the probability of longer-term recognition of more substantial policy failure. In all probability, by the time this longer-term recognition of policy failure is beginning to emerge, the government that initiated a particular policy intervention will have changed, … and the problem will be one for a subsequent government to deal with. Such subsequent governments can then simply disown or analyse away the deficiencies of old policy approaches, and start the whole process over again. (Sanders, 1991, p. 16)[1]

[1] A third aspect of my critique in 1991 was to call into question the 'desirability' and 'appropriateness' of pursuing 'statistical equality' between Aboriginal and other Australians as a policy goal. This possibly had echoes of the disowned 1960s policy of 'assimilation' in Australian Indigenous affairs, and sat somewhat uncomfortably with the 'self-determination' and 'self-management' policies of the 1970s and 1980s. This was particularly so for Indigenous people in remote areas who in the 1970s and 1980s were engaged in the 'homelands' movement (Sanders, 1991, p. 17).

This echoes Edelman's analysis in the 1970s and anticipates the 2010 debate over a framework for establishing policy success in which Marsh, McConnell and Bovens (2010) engaged. It also sits comfortably with the third type of account of policy and governing developed by Colebatch and Bacchi, in which problem construction is recognised as contingent and changeable in time and place. In 1991, I was already showing an inclination towards constructivist or interpretivist policy analysis, unwilling to settle easily on a single, clear idea of policy success or failure.

Back in 1991, I could also have argued that the announcement of the AEDP in 1987 had been a significant success for the Hawke government in Indigenous policy. From 1983 to 1986, the Hawke government's political management of Indigenous policy had been disastrous, as they slowly dishonoured a 1983 election promise of national land rights legislation, and in the process abolished the elected National Aboriginal Conference (Altman & Dillon, 1985; Jennett, 1988). Under a new minister in 1987, political management of Indigenous policy began to improve. A new representative body of Indigenous people was foreshadowed, in the form of a Commission that would share executive decision-making in the portfolio with the minister (Hand, 1987). Plus, a Royal Commission into Aboriginal Deaths in Custody was announced late in the year. While these two initiatives announced during 1987 promised future improvement after consultation and inquiry processes, the AEDP offered immediate expansion of programs for Indigenous employment, including the popular CDEP. This was a successful bundle of initiatives for the new minister in 1987, establishing improved relations with Indigenous people which lasted for the next two years. AEDP made a significant contribution to the successful political management of Indigenous policy between 1987 and 1989, while other initiatives like ATSIC and the Royal Commission, were progressing more slowly. My predictable failure analysis of AEDP was, thus, in the long-term programmatic mode of policy analysis, looking at that one policy in isolation and in relation to stated, rather unrealistic, policy goals. In broader political management, the AEDP had already been a great success in the years between when it was announced and my writing about its predictable failure in 1991.

## 3.4 A Convenient Change of Discourse: Failure and Change Under the Late Howard Governments

My second major encounter with failure analysis in Australian Indigenous affairs began in 2004, when Prime Minister Howard announced in April that the Coalition government intended to abolish the Aboriginal and Torres Strait Islander Commission (ATSIC), and then proceeded in November to be re-elected for a fourth consecutive term. Howard had never been a supporter of ATSIC, having opposed its establishment by the Hawke government when opposition leader in 1989. Yet, in government from 1996, Howard had allowed ATSIC to continue. Over the next eight years, ATSIC's leaders had developed its role in some innovative ways. But come 2004, a couple of ATSIC's leaders were facing political (and legal) troubles, and the Labor opposition, led by Mark Latham, announced it would abolish ATSIC if elected (Latham & O'Brien, 2004). Howard seized the opportunity to match Labor and be rid of ATSIC. He did so by declaring ATSIC a 'failure', both as an 'experiment in separate representation' and as an administrative arrangement for Indigenous-specific programs (Howard & Vanstone, 2004). Not only was ATSIC to be abolished as a national elected Indigenous representative body by amendment of legislation, but its program budgets were immediately dispersed ('mainstreamed') to line departments of government from July 2004.

Latham and Howard's moves in 2004 struck me as a convenient change of discourse around ATSIC, which combined labelling ATSIC as a failure with a major push for institutional change. My reaction was to defend ATSIC, pointing to its achievements and strengths, including 'distinctive, appropriate programs' and its encouragement of the 'political participation of Indigenous people' (Sanders, 2004). Like the official reviewers of ATSIC in 2003, I argued for the continued relevance of the principles and ideas behind ATSIC, and for its reform rather than abolition (Hannaford et al., 2003). This defence of ATSIC as a success worth keeping and reforming, was however against the tide of public policy debate. Legislation completing the abolition of ATSIC as a national elected Indigenous representative body was passed through parliament in early 2005.[2]

During their fourth electoral term, the Howard Coalition government proceeded to repeat and entrench this 'failure and change' style of analysis in Indigenous policy. From early 2006, a new minister for Indigenous affairs, Mal Brough, increased the denigration of past policy and combined it with a 'crash through', 'take no prisoners' military style. Major distinctive programs inherited from ATSIC by line departments, like the Community Housing and Infrastructure Program (CHIP) and CDEP, were reviewed, declared failures, and massively changed (Department of Families, Community Services & Indigenous Affairs, 2007). An inquiry into child abuse in the Northern Territory diagnosed a failed child welfare system and was used by Howard and Brough to launch the Northern Territory Emergency Response (NTER), a raft of policy measures that included more police, tighter alcohol restrictions, and income management of social security payments (Altman & Hinkson, 2007, pp. 1–2).

This intensifying barrage of failure analysis and the associated calls for institutional and policy change were, I argued, 'a generational revolution' in Australian Indigenous policy. This involved the 'moral denigration of the previous generation of people working in Indigenous affairs' and an attempt to build 'a new institutional architecture and policy rhetoric for the next generation's efforts' (Sanders, 2008, pp. 197, 201). I argued that similar moral dynamics had led to an earlier generational revolution in Australian Indigenous policy in the decade following the famous 1967 constitutional alteration referendum, when the Commonwealth became involved as

---

[2] One regional elected Indigenous representative body survived these legislative changes, the Torres Strait Regional Authority. More Torres Strait Islanders now live outside the region than in it, however, and the representative body for this diaspora, the Torres Strait Islanders Advisory Board was abolished along with the rest of ATSIC. How the TSRA avoided the 'failure' label is, to my knowledge, an untold story.

a major *national* player in Indigenous policy (rather than just in its Territories). In that instance, it was State and Territory Aboriginal welfare authorities who were seen as failing, while the new Commonwealth Indigenous affairs architecture, developed particularly under the Whitlam Labor government from 1972, was to be the new and better way forward. This comparison of Howard and Whitlam was something of a challenge, both to myself and to others who regarded these two prime ministers very differently in relation to Indigenous policy.

## 3.5 Stories of Success: An Unexpected Reaction

Early in 2007 I attended a conference in Melbourne and heard Bill Moss, formerly of Macquarie Bank, tell a success story about engagement with Indigenous Australia as a philanthropist in retirement. With others, Moss had connected with Titjikala community, south of Alice Springs, and helped it establish an upmarket tourist camping venture. Moss and his associates supplied the upmarket tent accommodation and Titjikala provided the authentic Indigenous community experience when tourists turned up. This was presented as a simple small-project success story, a win–win for the community and for tourists. It was also presented as relatively easily done and, in principle, replicable as general policy success.

My reaction to this success story was unexpected. While I had become critical of the emphasis on failure in Indigenous policy during the late Howard governments, this success story also made me uneasy. Was there something shared by these narratives of simple small-project success and big policy failure?

While I had not done research at Titjikala directly, I knew that they had enjoyed stable community government administration for a decade, with a dedicated and innovative manager of CDEP.[3] On my next research trip to central Australia, I met with this CDEP manager in an attempt to understand the tourism venture success story. It was, as I suspected, built on CDEP, as most things in Titjikala then were. Community members provided experiences for the tourists as part of their morning's work on CDEP, or as work for top-up wages if done after hours. This was a success story, without a doubt, but a fragile one. The tourism camping facilities were just a piece of capital investment, with no provision for recurrent costs. These were borne by the community through CDEP, with income from tourist accommodation adding top-up to CDEP wages.

The fragility of this success story became evident from August 2007 when the Commonwealth Department of Employment and Workplace Relations began closing CDEPs south of Alice Springs as its contribution to the NTER. Without CDEP to

[3] See Harry Scott. 'Job creation and 'mutual obligation': Tapatjatjaka Community Government Council, Northern Territory', in Morphy and Sanders (2001, pp. 207–208).

build on, Titjikala's tourist camping venture struggled and faded within a year.[4] Observing this change of fortunes, I began to think harder about what big policy failure narratives and small-project success stories have in common, and why I found both unconvincing. This led me to the idea of two rhetorical registers and the different roles they play in the moral dynamics of Australian Indigenous policy.

## 3.6 Challenging and Calming Rhetorical Registers: New and Established Participants

Big policy failure narratives and small-project success stories are both part of a challenging rhetorical register in Australian Indigenous policy. This register challenges participants in Indigenous policy to improve on the past. Talk of big policy failure condemns the past as inadequate, while stories of small-project success show what is possible in the present and future with good will and imagination. Within the moral dynamics of Indigenous policy, this is a rhetorical register that aims for substantial improvement in the future, with hope and optimism.

I label the alternative the calming rhetorical register. This can be the register of long-established participants in Indigenous policy, and by the early 2000s after two decades of academic involvement, I was certainly one of these. This register tends to emphasise the importance of long-term commitment in Indigenous affairs, arguing that enduring success can be harder to achieve and more complex than is often admitted in short-term success stories—as the fading of the Titjikala tourism venture unfortunately demonstrated. Plus, the calming register may not be so dismissive and disparaging of the past, partly because some who use it are themselves part of that past.

In early 2007, I looked around for others who seemed uncomfortable with the rise of the challenging register under the late Howard governments and were, like me, operating in the calming register. Two participants adopting a calming rhetorical style came to my attention. One was the Chief Minister of the Northern Territory, Clare Martin, and the other was my colleague at the Australian National University (ANU), David Martin. I deal with each separately, as they articulated their calming rhetoric somewhat differently.

In responding during 2006 to the Howard government's new combative minister for Indigenous affairs, Clare Martin was a model of measured underreaction. As Chief Minister of the Northern Territory since 2001, and Labor opposition leader since 1999, Martin did not like being told what was needed by a brash new Commonwealth Coalition minister. Martin gave a significant speech in Canberra in May 2006 entitled 'Eyes Wide Open'. She began by acknowledging that the 'state of Aboriginal communities across the Territory—indeed across the country—is a matter of

[4] The Tapatjatjaka Community Government Council which ran this CDEP also disappeared in 2008, amalgamated into the new and much larger MacDonnell Shire as part of the Northern Territory Labor government's gran plan for local government reform. See Sanders (2013).

national significance' and that 'it requires urgent national attention ... and a long-term commitment'. However, Martin continued by saying that, if the 'recent debate' about Aboriginal communities had 'shocked many Australians', it had come as 'no great surprise to Territorians':

> We have our eyes wide open .... And we know the problems that many of our communities face. (Martin, 2006a, pp. 2–3)

This was the moral positioning of an old-hand, who had been in the thick of Indigenous issues in Australian public policy for a decade. Martin went on to call for higher levels of 'long term investment in infrastructure, education and employment' by the national government, as the Territory government was already 'stretched to the limit'. The remainder of Martin's speech was essentially a catalogue of cases in which 'unsung heroes' were 'tackling the issues' and 'making some headway', but in a context of inadequate resources (Martin, 2006a, pp. 5–6).[5]

In this calming rhetorical register, Martin's success stories were partial, complex, and fragile. They were built on the involvement of Aboriginal people, locally, as well the resources and policy support of governments and corporations, regionally. As Martin put it towards the end of her speech:

> So, some really good things are happening on the ground – but the Government also has to play its part.

The NT Chief Minister's final call was for 'a 20 year plan—a 20 year commitment', in order to 'get out of this cycle of Aboriginal issues being at the top of the agenda one day and forgotten the next' (Martin, 2006a, pp. 14–18).

My ANU colleague David Martin was more willing to accept the language of policy 'failure' in 2006, but cautioned that the 'new direction' in Indigenous policy was as likely to 'fail' as the 'old' because it similarly lacked engagement with the 'agency of Aboriginal people'. Drawing on family experience as well as academic field work, Martin depicted a 'moral economy' among Indigenous people in regional and remote areas which is a 'fundamental structure of kinship' reaching well beyond the 'private realm' into 'economic and social and political relations' (Martin, 2006b, passim). The implication was that 'market–based social policy frameworks' would have little power in these circumstances and that far greater engagement was needed with 'Aboriginal organisations, as mediating and transforming institutions' (Martin, 2006b, pp. 8–9).

## 3.7 Established Participants and the Challenging Register: More Complex Dynamics

In late 2007, three events made me think more about rhetorical registers in Indigenous policy: a lecture, a change of Northern Territory Chief Minister, and the publication of a book. I deal with each in this order.

---

[5] The Titjikala tourism venture was one of these success stories.

The first, in October, was the seventh annual Charles Perkins memorial lecture, delivered at the University of Sydney by one of the six Indigenous Labor members of the Northern Territory Legislative Assembly, minister Marion Scrymgour. While part of Clare Martin's Territory Labor government, this Indigenous minister was *not* operating in the calming rhetorical register.

Starting with a brief account of her father's removal from his family in central Australia in the 1930s as part of an 'earlier Northern Territory Intervention' by the Commonwealth, Scrymgour then turned her attention to the 'second Intervention' that had started in June 2007. She was scathing of Howard's use of the idea of a national emergency response and accused him instead of pulling a 'rabbit out of a hat' for the upcoming election, 'the black kids' Tampa' as she called it, referencing Howard's tactics back in the 2001 election (Scrymgour, 2007, p. 5). Scrymgour argued that:

> Aboriginal women had been begging for action from Howard over a raft of social problems the best part of a decade – entreaties which he had studiously ignored or paid lip service to. (Scrymgour, 2007, p. 5)

What Howard and Brough were now offering, through 500 + pages of emergency response legislation, bore 'no relationship', she argued, to the recommendations of the *Little Children are Sacred* report, to which they were purportedly responding. Rather, Scrymgour argued:

> Aboriginal Territorians are being herded back to the primitivism of assimilation and the days of native welfare. It has been a deliberate, savage attack on the sanctity of Aboriginal family life. (Scrymgour, 2007, p. 6)

Noting Brough's repeated attacks on 'anyone who has raised doubts and fears about this new world order for Aboriginal Territorians', Scrymgour reached a rhetorical high-point in her lecture, thus:

> John Howard and Malcolm Brough, this evening I am doing far more than merely criticising you and your government's assault on Aboriginal Territorians, I am condemning its motivation; I am condemning its operations; and I am condemning – outright – its moral basis and the moral authority you purport to exercise in 'saving the children'.
>
> You are doing nothing of the kind.
>
> And as someone who has worked in Aboriginal health for much of my working life; as someone who has focussed on child protection, and indeed as someone who now has ministerial responsibility for child protection, I will not be cowed by the bully boy tactics of McCarthyism – old or new. (Scrymgour, 2007, pp. 6–7)

In the second half of her lecture Scrymgour developed the theme that an 'impending social crisis' had been slowly building in remote Indigenous communities over many years. She also developed a second crisis theme, about Australian 'settler society' lacking the 'capacity to abandon past thinking about colonialism' and being unable to 'imagine other ways of ordering the world' (Scrymgour, 2007, pp. 7–13). This was the second, more real and fundamental national emergency in Scrymgour's view, captured in her lecture title as covering 'Caboolture and Kirribilli' rather than 'Milikapiti and Mutitjulu'.

Towards the end of her lecture, Scrymgour offered one example of a success story, in western Arnhem Land, where Aboriginal families and scientists were 'harnessing traditional burning practices to be utilised as greenhouse gas off sets for the gas industry' (Scrymgour, 2007, p. 14). This was a brief glimpse of the calming rhetorical register, but most of Scrymgour's lecture was in the challenging rhetorical register. Indeed, while playing the old-hand card as someone with long involvement in Indigenous health, Scrymgour was *doubly* in the challenging register. Rather than using calming rhetoric, like her Chief Minister boss, minister Srymgour was challenging the challengers, Howard and Brough, in their own challenging style of rhetoric.

The second event which made me pause for thought, in November 2007, was the ousting of Clare Martin as Northern Territory Chief Minister, after six years in the job. Media reports suggested that insurrection within the ranks of Territory Labor had started back in 2006, when Martin was calmly responding to Brough's challenging rhetoric, and that this insurrection had resurfaced during the 2007 Federal election campaign, when Rudd wanted Martin to keep quiet and not criticise the NTER (Rothwell, 2007). Within days of Rudd's election, a new Chief Minister of Northern Territory emerged, Paul Henderson, who was more on Rudd's wavelength than Clare Martin in accommodating most of the NTER. But Henderson's new Deputy Chief Minister was Territory Labor's most strident critic of the NTER, Marion Scrymgour.

This combination, of Scrymgour ascending with doubly challenging rhetoric and Martin descending with calming rhetoric, made me wonder whether calming rhetoric can ever really work for politicians on a long term basis. My newly formulated question was: does the drive to improvement in public policy always push politicians towards the challenging rhetorical register in order to survive? Perhaps it is only securely employed academics and bureaucrats who can afford the luxury of operating in the calming rhetorical register and keeping their jobs.

The third event in late 2007 that intrigued me was the publication of a book in which two long-established Indigenous policy bureaucrats also adopted the challenging register. *Beyond Humbug: Transforming government engagement with Indigenous Australia* (2007) was written by two men who had been deeply involved in Indigenous policy since the 1970s, Michael Dillon and Neil Westbury. Between them they had occupied numerous senior positions in ministerial offices, the Department of Aboriginal Affairs, ATSIC, land councils, and the Office of Indigenous Policy within the Department of Prime Minister and Cabinet. In the 1990s, they had significant roles in shaping the *Native Title Act* as the Commonwealth's response to the 1992 High Court *Mabo* decision. In the early 2000s, both had worked in senior executive positions in Darwin for the Northern Territory government. Yet their 2007 book developed as its major argument that government had become *disengaged* from Indigenous Australia during the previous three decades, particularly in remote areas.

I found this central argument of Dillon and Westbury's book somewhat incredible, as it strained against their own deep involvement in Indigenous issues as government officials over multiple decades. But their argument worked by and because it juxtaposed an analysis of policy and program failure with a call for fundamental change. The current 'Architecture of Government' and 'Program Framework in Indigenous

Affairs' had failed, in the view of Dillon and Westbury (2007, pp. 61–65). The 'overlapping mosaic' of Indigenous-specific programs and general programs spread across many government agencies had failed to address deep Indigenous disadvantage and was beset by 'implementation challenges' (Dillon & Westbury, 2007, pp. 50–78). The 'way forward' for Dillon and Westbury was, in one iteration, a seven-point list of reforms that re-emphasised regionalism (as ATSIC had done), but also turned away from Indigenous-specific programs (because they were under-resourced and provided second-rate services) in favour of 'mainstream programs' run by better resourced line agencies on 'needs-based criteria' (Dillon & Westbury, 2007, pp. 78–81). Another iteration of their way forward was getting rid of 'parallel systems of service delivery' and reforming 'fiscal federalism' in the hope of overcoming 'institutional disengagement by government' and 'acknowledging Indigenous cultural perspectives' (Dillon & Westbury, 2007, pp. 197–198).

My reaction to Dillon and Westbury's ideas for institutional reform was that they were divided between the opportunistic and the unrealistic. On the one hand, Dillon and Westbury seemed to agree opportunistically with the Howard government's move away from Indigenous-specific programs in an Indigenous-specific organisation (ATSIC), towards gaining more for Indigenous people through 'mainstream'[6] programs administered by line departments of government. On the other, Dillon and Westbury were seeking fundamental reform of policy institutions deeply embedded in Australian federalism, around fiscal relations, and the chance of these changing due to Indigenous policy seemed to me miniscule. The fundamental reform aspect of Dillon and Westbury's ideas, along with their emphasis on past policy failure, put them squarely in the challenging rhetorical register. Despite their deep involvement in it, past policy was treated as predominantly mistaken, and the future was envisaged as something that could be so much better, with a combination of political will and reasoned analysis.[7]

During 2006 and 2007, it seemed to me, established participants in Australian Indigenous policy were becoming divided. Some wanted to defend the past in which they had been involved, and to recognise its achievements like ATSIC, CDEP, and CHIP. Others, by contrast, were more willing to disown the past and embrace the generational revolution in Indigenous policy in the hope of gaining more resources and doing things better in the future. While the challengers embraced the idea of past policy failure and hoped for a better future through big reforms, the calmers were more likely to point to existing successes that should be preserved and built upon, particularly in Indigenous community-based organisations.[8]

---

[6] I prefer to call these general programs, but the language of 'mainstream' programs, and mainstreaming, is deeply embedded in Indigenous policy debates.

[7] This basic structure of argument has been repeated twelve years on in Westbury and Dillon (2019). I have responded with a range of criticisms in Sanders (2021).

[8] See, for example, Australian Collaboration (2007). This handbook compiled 16 instances of successful Indigenous organisations. Other examples of success in Indigenous organisations were documented from 2005 through the Indigenous Governance Awards run by Reconciliation Australia and BHP Billiton.

## 3.8 Beyond Predictable Failure in Responding to Rudd's Closing the Gap: Competing Principles and Alternative Idioms

The Rudd Labor government elected in November 2007 distinguished itself from the Howard Coalition government of the previous 11 years by apologising for past policies of Aboriginal and Torres Strait Islander child separation, as recommended in the *Bringing them home* report a decade previously (Human Rights & Equal Opportunity Commission, 1997). In a contrary vein, however, Rudd built on the failure analysis and the Indigenous policy revolution of the late Howard years (Sanders & Hunt, 2010). ATSIC was now dead, and thoroughly disowned by Labor, so work on a new national Indigenous representative body proceeded slowly and in new institutional directions. Also, Rudd only lightly reworked Howard's 2007 Northern Territory Emergency Response/Intervention. This reworking included a 12 month moratorium on closure of CDEPs in the Northern Territory, and even some temporary reinstatements during 2008. But in the longer term, from 2009, the failure framing of CDEP prevailed and, like ATSIC, it was abolished.

The Rudd government's new centrepiece Indigenous policy was a Closing the Gap strategy focused 'on life expectancy, educational achievement and employment opportunities', first announced in the latter part of Rudd's apology speech to Parliament in February 2008. 'Concrete targets for the future' were set, aiming to halve identified statistical gaps 'within a decade' or eliminate them 'within a generation'. Existing approaches in Indigenous affairs were labelled as 'business as usual' and 'not working'. What was needed was:

> A new beginning which contains real measures of policy success or policy failure; a new beginning, a new partnership, on closing the gap with sufficient flexibility not to insist on a one-size-fits-all approach for each of the hundreds of remote and regional Indigenous communities across the country but instead allowing flexible, tailored, local approaches to achieve commonly-agreed national objectives that lie at the core of our proposed new partnership. (Rudd, 2008)

Rudd was clearly operating in the challenging rhetorical register, condemning past policy, and envisaging something new and better for the future.

Rudd's Minister for Indigenous affairs, Jenny Macklin, also operated in the challenging rhetorical register when she addressed the National Press Club in late February 2008.

> Decades of failure have spawned, inside our national boundary, a country within a country. I've seen it. And it's not just another country, it's the unluckiest of countries – characterised by neglect, disadvantage and poverty…
>
> This must not go on. We must find new ways of doing things because the old ways have so comprehensively failed. In doing so, we must work with Indigenous Australians in a partnership built on respect and mutual responsibility.
>
> Inevitably there will be difficult decisions but all these decisions will be driven by one single criterion – evidence. This is the Government's obsession and we make no excuses for it. It is my abiding fixation and I readily acknowledge it.

> All our policy decision-making will be based on a thorough, forensic analysis of all the facts and all the evidence. Once implemented, all programs will be rigorously and regularly evaluated. This is the principle I will impose on my portfolio. (Macklin, 2008, pp. 3–4)

One of my ANU colleagues, Janet Hunt, responded immediately to Macklin, criticising her emphasis on failure and the questionable logic of future policy combining 'new ideas' and 'evidence':

> So Macklin is looking for 'new ideas' about how to tackle the apparently intractable problems facing indigenous communities.
>
> As I sat listening to her I wondered how people working their butts off in indigenous organisations would feel about what she was saying. Was she consigning all their work to the failure bin too? What exactly has failed? And more importantly, what has succeeded?...
>
> The Australian Collaboration has documented a series of indigenous success stories in organisational and business achievement. Indigenous musicians, artists, sports people, lawyers, educators and many more can hold their own against the world. But we continue to focus on failure.
>
> Macklin says she wants new ideas, but she also wants to proceed only where there is an evidence base. New ideas, by their very nature, have not been tried, so there is no evidence about whether they will work or not….
>
> I am not against new ideas, but to write off the past and what is happening now is to miss the evidence Macklin is looking for. (Hunt, 2008a; see also Hunt, 2008b)

Here was another established player trying to use the calming rhetorical register, against the overwhelming tide of challenging rhetoric in Indigenous policy. Past policy was seen as involving success as well as failure, and the challenge for the present was in knowing what to keep from the past, and what to support into the future, as well as developing new ideas.

My own reaction to the Rudd government's embrace of failure rhetoric in Indigenous affairs was slower to emerge than Hunt's, and in the end I focused more on the evidence half of the new government's rhetorical construction.[9] In the Rudd government's strong use of the sacred authoritative choice account of policy and governing, evidence was opposed to ideology in a good/bad dichotomy. I questioned this dichotomy by pointing to three competing principles in Indigenous policy: equality, choice, and guardianship, and also to three interpretations of the dominant equality principle (Sanders, 2010). When evidence was cited, which of the principles or interpretations was it in relation to? Arguments needed to be constructed with data, rather than evidence speaking for itself, I argued, following a school of thought that had flourished in America and elsewhere since the late 1980s (Majone, 1989; Stone, 1988; Yanow, 2000).[10] Australian policy history also suggested that balancing these competing principles in Indigenous affairs was forever-changing, which defied simple unitary assessments of policy success or failure. Further, I argued, ideologies were not all bad, as they represented genuinely held, different political philosophies, that rightly should be brought to bear in Indigenous policy debates. My sense of

---

[9] The reaction of my 1990 encourager and collaborator, Jon Altman, who had since become my supervisor at the ANU, emphasised the 'valuing of diversity' in Indigenous. See Altman (2009).

[10] See also Maddison (2012) for another questioning of an evidence-only approach.

Indigenous policy as an ongoing debate, and a balancing act, drew more from Colebatch's second account of policy and governing rather than the dominant, and sacred, authoritative choice account used so predominantly within government.

In time, drawing on Rowse (2012), I added to this analysis of Indigenous policy the idea of two alternative idioms (Sanders, 2014). On the one hand, there is a 'populations' idiom, which uses socioeconomic statistics to identify 'disadvantage' among Indigenous individuals, households and regions and to build cases for remediating policies. On the other hand, there is a 'peoples' idiom, which sees Indigenous political communities as rights-bearing agents already organised and active in Australian politics and internationally. As Rowse (2012, p. xv) noted, these alternative idioms involve very different ways of recognising Indigenous people, and very different notions of social justice, as pursued through Indigenous policy. As Rowse put it:

> The dissonance between these recognitions and these two conceptions of social justice animates and at times confuses the rhetorical landscape of contemporary Indigenous affairs policy. (Rowse, 2012, p. xv)

Alternative idioms cannot be balanced in Indigenous policy debates, like the three competing principles. They are different social constructs which must be chosen or moved between. It is almost impossible-to-conduct debate in both idioms simultaneously. Identifying alternative idioms is analysis of Indigenous policy in a third type of account of policy and governing.

My analysis of Indigenous policy was becoming more complex the longer I watched and thought, possibly to the chagrin of some of my students and colleagues. Yet some warmed to this multi-dimensional complexity. If Indigenous policy moves between alternative idioms, while also balancing three competing principles and three interpretations of the dominant equality principle, then criteria are proliferating against which to judge policy a success or failure. Indeed, policy is highly likely to be simultaneously succeeding against one criteria, while failing against another. Indigenous policy does not so much succeed or fail, as succeed and fail at the same time in different dimensions.

## 3.9 Concluding Comments

These concluding remarks deal briefly again with policy theory, and with the predominance of ideas of failure over ideas of success in policy making. Policy theory has progressed significantly during my four decades as an academic political scientist. As an undergraduate, I was exposed to a debate between rational decision making and incrementalism, with the former seen as desirable normative theory and the latter as realist empirical practice (see McGrew & Wilson, 1982 for a collection of contributions to this debate). Colebatch's work has built from this debate, while changing its terms considerably. His authoritative choice account has its roots in ideas of rational decision making, but also in ideas of the policy prerogatives of democratically elected politicians, whether based in reasoned analysis or political philosophy/ideology. His structured interaction account resonates with incrementalism, which always contained an alternative normativity as well as being better at describing government. The greatest progress in policy theory over these decades, however, has been in the development of the third social construction account of policy and governing. That problems are constructed in policy processes at the same time as being addressed, is an insight of great importance, that adds a third dimension to our understanding of governing. Lukes (1974) offered a three-dimensional analysis of the concept 'power' in the 1970s, and by the 1980s this insight was beginning to filter into debates about policy making. Bacchi has been our Australian standard bearer in this broader international move towards thinking about framing and problematisation in policy processes. Meanwhile, Colebatch has provided a masterly synthesis and overview of all three dimensions of policy and governing, through his three accounts approach. My use of Colebatch's and Bacchi's work to understand Australian Indigenous policy hopefully shows that this is theory of great practical application, in a policy arena of great national importance and strong moral commitments. My addition of improvement as a fourth forceful theme in policy, alongside Colebatch's authority, order, and expertise, has been done specifically in the context of teaching and researching Indigenous policy. But perhaps the idea of improvement is usefully recognised as at the heart of *all* policy making and analysis.

Focussing specifically on ideas of success and failure, this enhanced policy theory allows us to see that there are multiple rationalities and dimensions in which assessments can be made. The idea of failure predominates over success because it is used by participants in the moral drive to improve government action over time, as part of the challenging rhetorical register. Small success stories can also be part of this register, to indicate possible futures for larger-scale policy improvement. But the more systematic use of the idea of past policy success is restricted to the much more limited number of participants who operate in the calming rhetorical register.

Indigenous policy in Australia is now more than a decade on from Rudd's Closing the Gap policy. Four prime ministers since have continued the annual report back to parliament each February that Rudd initiated. This itself could be seen as a success of political management, having established a process that keeps Indigenous policy on the parliamentary agenda at least once each year. However, at the more programmatic level, as targets have not been met, Closing the Gap has been through its inevitable ritual evisceration as a failure (see for e.g. Oscar & Little, 2018). Under Morrison as Prime Minister since late 2018, a Coalition of Aboriginal and Torres Strait Islander Peak Organisations has contributed to new 'Closing the Gap in Partnership' targets for the decade ahead.[11] The signing of an agreed document in July 2020 was presented by the Indigenous leaders involved as a success, a move towards shared decision-making.[12] But other Indigenous people called the new policy and targets mere 'buzzwords' that are destined to fail (Watego, 2020). And they will in time be proven correct, as I was back in 2000.

---

[11] https://www.abc.net.au/news/2020-07-30/closing-gap-targets-agreement-aboriginal-torres-strait-islander/12506232.

[12] https://apo.org.au/node/307189.

Success and failure in public policy are strategic concepts used by participants as part of the processes of governing. They are asserted by different participants for very different strategic purposes, and it is perhaps naïve to think that there can ever be an objective, overarching assessment of either policy success or failure. Words often succeed in managing policy problems in the short term, only to be seen later to have fallen short in achieving programmatic outcomes. But it is almost the essence of government and public policy to aspire to improvement and later to fall short, only to aspire again and fall short again in never-ending repeated cycles. These are the moral and rhetorical dynamics of governing in general and of Indigenous policy more particularly.

In analysing 'disenchantment' with Australian Indigenous policy during the 2000s, Patrick Sullivan (2011, p. 87) observed that 'new policy' succeeds because it is 'future oriented' and can avoid scrutiny of its 'efficacy', while 'old policy' fails simply because it is old and not sufficiently oriented towards future improvement. This pithy comment encapsulates the cycles I have come to observe in Indigenous policy over four decades, of disowning past policy as a failure while brazenly reinventing something 'new and different' that is claimed will succeed. But in time the new and different becomes old, and it suffers that same old fate of being disowned. Is this as good as it gets? Or can we, by being analytically aware of these moral dynamics in policy and governing, possibly transcend them? This is my new question, which has grown out of repeated attempts to refine and answer those original, common-sense questions about success and failure in Australian Indigenous policy.

## References


Altman, J. (Ed.) (1991). Aboriginal Employment Equity by the year 2000, Research Monograph No. 2, Centre for Aboriginal Economic Policy Research, ANU, Canberra.

Altman, J. (2009). Beyond Closing the Gap: Valuing diversity in Indigenous Australia, Working Paper No 54, CAEPR ANU, Canberra.

Altman, J., & Dillon, M. (1985). Land rights: Why Hawke's model has no backing. *Australian Society, 4*(6), 26–29.

Altman, J., & Hinkson, M. (Eds.). (2007). *Coercive reconciliation: Stabilise, normalise, exit Aboriginal Australia.* Arena Publications.

Antonios, Z. (1997). The CDEP scheme and racial discrimination: A report by the race discrimination commissioner, *Human Rights and Equal Opportunity Commission*, Sydney, December.

Australian Collaboration. (2007). Maps to success: Successful Strategies in Indigenous Organisations, Australian Institute of Aboriginal and Torres Strait Islander Studies, Canberra.

Bacchi, C. (2009). *Analysing Policy: What's the problem represented to be*? Pearson Australia, Frenchs Forest NSW.

Bovens, M. (2010). A comment on Marsh and McConnell: Towards a framework for establishing policy success, *Public Administration, 88*(2), 584–585.

Colebatch, H. (2009). *Policy* (3rd ed.). Open University Press.

Department of Families, Community Services & Indigenous Affairs. (2007). Living in the Sunburnt Country – Indigenous Housing: Findings of the Review of the Community Housing and Infrastructure Programme, Price Waterhouse Coopers, February.

Dillon, M., & Westbury, N. (2007). *Beyond Humbug: Transforming indigenous engagement with indigenous Australia.* Seaview Press.

Edelman, M. (1972). *Politics as symbolic action: Mass Arousal & Quiescence.* Markham Publishing.

Edelman, M. (1977). *Political language: Words that succeed and policies that fail.* Academic Press.

Fischer, F., & Forester, J. (Eds.). (1993). *The argumentative turn in policy analysis and planning.* Duke University Press.

Gordon, Z. (2010). Deconstructing 'Aboriginal welfare dependency': Using postcolonial theory to reorient Indigenous affairs. *Journal of Australian Indigenous Issues, 14*(2), 14–29.

Hand, G. (1987). *Foundations for the Future: Policy Statement.* Australian Government Publishing Service, Canberra, July.

Hannaford, J., Huggins J., & Collins, B. (2003). In the hands of the Regions: A new ATSIC, Report of the Review of the Aboriginal and Torres Strait Islander Commission, Canberra.

Howard, J., & Vanstone, A. (2004). Transcript of the Prime Minister, Joint Press Conference with Senator Amanda Vanstone, Parliament House, Canberra, 15 April.

Human Rights and Equal Opportunity Commission. (1997). *Bringing them home: National Inquiry into the Separation of Aboriginal and Torres Strait Islander Children from Their Families.* Sterling Press.

Hunt, J. (2008a). Stress indigenous positives: The Rudd Government should forget the failures of the past and build on success stories, *Canberra Times*, 29 February.

Hunt, J. (2008b). Failure, Evidence & New Ideas. Topical Issue No 1/2008b, Centre for Aboriginal Economic Policy Research, ANU, Canberra. http://caepr.cass.anu.edu.au/sites/default/files/docs/Hunt_Ideas_0.pdf

Jennett, C. (1988). Aboriginal Affairs Policy. In C. Jennett & R. Stewart (Eds.), *Hawke and Australian Public Policy* (pp. 245–283). Macmillan.

Latham, M., & O'Brien, K. (2004). Opportunity and responsibility for indigenous Australians: Statement by Federal Labor Leader and Shadow Minister for Indigenous Affairs, Canberra, 30 March.

Lukes, S. (1974). *Power: A radical view.* The Macmillan Press.

Macklin, J. (2008). Closing the Gap – Building an Indigenous Future, Minister for Families, Housing, Community services & Indigenous Affairs, Address to the National Press Club, 27 February.

Maddison, S. (2012). Evidence and contestation in the indigenous policy domain: Voice, ideology and institutional inequality. *Australian Journal of Public Administration, 71*(3), 269–277.

Majone, G. (1989). *Evidence, argument and persuasion in the policy process*, BookCrafters, Chelsea Michigan.

Marsh, D., & MConnell, A. (2010). Towards a framework for establishing policy success, *Public Administration, 88*(2): 564–583.

Martin, C. (2006a). 'Eyes Wide Open': Keynote Address of Aboriginal Affairs, 7th Annual Fraser Lecture, Belconnen Labor Club, Canberra, 31 May.

Martin, D. (2006b). Why the 'new direction' in Federal Indigenous affairs policy is as likely to 'fail' as the olds directions, Topical Issue No.5/2006b, CAEPR ANU, Canberra. https://caepr.cass.anu.edu.au/research/publications/why-new-direction-federal-indigenous-affairs-policy-likely-fail-old-directions

McGrew, A. G., & Wilson, M. J. (1982). *Decision making: Approaches and analysis.* Manchester University Press in association with the Open University.

Morphy, F., & Sanders, W. (Eds.) (2001). *The indigenous welfare economy and the CDEP scheme*, Research Monograph 20, CAEPR ANU, Canberra.

Oscar J., & Little, R. (2018). Time to reset Closing the Gap: 'policy paralysis is not an option', *The Mandarin*, 8 February. https://www.themandarin.com.au/88187-closing-the-gap-urgent-need-to-reset/

Rothwell, N. (2007). Clare on the back foot, *The Weekend Australian*, 27 October.

Rowse, T. (2012). *Rethinking social justice: From 'peoples' to 'populations.'* Aboriginal Studies Press.

Rudd, K. (2008). Apology to Australia's Indigenous Peoples, Commonwealth of Australia, Parliament Debates: House of Representatives.13 February.
Sanders, W. (1985). The Politics of Unemployment Benefits for Aborigines: Some Consequences of Economic Marginalisation, in D Wade-Marshall & P Loveday eds Employment & Unemployment: A collection of papers, ANU North Australia Research Unit, Darwin, pp. 137–162.
Sanders, W. (1988). The CDEP scheme: Bureaucratic politics, remote community politics and the development of an aboriginal 'Workfare' program in times of rising unemployment. *Politics, 23*(1), 32–47.
Sanders, W. (1991). Destined to fail: The Hawke government's pursuit of statistical equality in employment and income status between Aborigines and other Australians by the year 2000 (or, a cautionary tale involving the new managerialism and social justice strategies. *Australian Aboriginal Studies, 1991*(2), 13–18.
Sanders, W. (1993). The rise and rise of the CDEP scheme: an Aboriginal 'workfare' program in times of persistent unemployment. Discussion Paper No. 54, CAEPR ANU, Canberra.
Sanders, W. (1997). How Does (and Should) DSS treat CDEP participants? (What are these allegations of racial discrimination?), *Social Security Journal*, December 1997, 19–30. Also available as Discussion Paper No. 149, CAEPR ANU, Canberra.
Sanders, W. (2004). ATSIC's achievements and strengths: Implications for Institutional Reform, *Journal of Australian Indigenous Issues, 7*(3) 2004: 14–21. Also available as Topical Issue No. 10/2004, CAEPR, ANU, Canberra. https://caepr.cass.anu.edu.au/research/publications/atsics-achievements-and-strengths-implications-institutional-reform-atsics
Sanders, W. (2008). In the name of failure: A generational revolution in Indigenous affairs. In C. Aulich & R. Wettenhall (Eds.), *Howard's Fourth Government: Australian Commonwealth Administration 2004–2007* (pp. 187–205). UNSW Press.
Sanders, W. (2010). Ideology, Evidence and Competing Principles in Australian Indigenous Affairs: From Brough to Rudd via Pearson and the NTER, *Australian Journal of Social Issues, 45*(3): 307–331. Also available as Discussion Paper No. 289/2009, CAEPR ANU, Canberra.
Sanders, W. (2013). Losing localism, constraining councillors: Why the Northern Territory supershires are struggling. *Policy Studies, 34*(4), 474–490.
Sanders, W. (2014). Experimental governance in Australian Indigenous affairs: From Coombs to Pearson via Rowse and the competing principles. Discussion Paper No. 291, CAEPR ANU, Canberra.
Sanders, W. (2016). 'Reframed as Welfare: CDEP's fall from favour' in Kirrily Jordan ed *Better than welfare? Work and Livelihoods for Indigenous Australians after CDEP*, CAEPR Research Monograph No. 36, ANU Press, Canberra, 2016, 31–49.
Sanders, W. (2017). Three accounts of the emergence of the remote jobs and communities program: Changing timeframes and types of actors. *Australian Journal of Political Science, 52*(2), 272–287.
Sanders, W. (2021). Big Picture Essays on Australian Indigenous Policy: Deep Structures and Decolonising, Policy Insights Special Series Paper No.5, CAEPR ANU, Canberra.
Sanders, W., & Hunt, J. (2010). Sorry, but the Indigenous affairs revolution continues, in C Aulich & M Evans eds *The Rudd Government: Australian Commonwealth Administration 2007–2010*, ANU E Press, Canberra.
Scrymgour, M. (2007). Whose national emergency? Caboolture and Kirribilli? Or Milikapiti and Mutitjulu? Charles Perkins Oration, University of Sydney, 23 October.
Spicer, I. (1997). *Independent Review of the Community Development Employment Projects (CDEP) Scheme*. Office of Public Affairs.
Stone, D. A. (1988). *Policy paradox and political reason*. Scott.
Sullivan, P. (2011). *Belonging together: Dealing with the politics of disenchantment in Australian Indigenous Policy*. Aboriginal Studies Press.
Watego, C. (2020). The 'new' Closing the Gap is about buzzwords, not genuine change for Indigenous Australia. *The Conversation*, 31 July. https://theconversation.com/chelsea-bond-the-new-closing-the-gap-is-about-buzzwords-not-genuine-change-for-indigenous-australia-143681

Westbury, N., & Dillon, M. (2019). Overcoming Indigenous Exclusion: Very hard, plenty humbug, Policy Insights Special Series Paper No.1, CAEPR ANU, Canberra.
Yanow, D. (2000). *Conducting interpretive policy analysis.* Sage Publications.

# Chapter 4
# Caring for Country as Deliberative Policymaking

**Justin McCaul**

**Abstract** Since the historic *Mabo* judgment and the recognition of native title, Aboriginal and Torres Strait Islander peoples have sought to reclaim their traditional lands and waters and reassert political authority and management responsibility over their cultural heritage and natural resources. This practice, referred to as 'Caring for Country', has evolved over time to become a prominent area of Indigenous public policy. As I will argue in this chapter, Caring for Country is an example of Indigenous led policy expertise and success that contrasts sharply with the decades of government administered Indigenous policy failure. Behind this success has been Aboriginal and Torres Strait Islander peoples' belief that Caring for Country is not simply land management, but an exercise in Indigenous self-government. Drawing on deliberative democracy theory, I will argue that Caring for Country can be understood as a form of 'deliberative policymaking' in which Aboriginal and Torres Strait Islander people challenges state centred policymaking that expects Indigenous people to be passive recipients of state policy.

**Keywords** Caring for country · Deliberative democracy · Indigenous public policy · Native title · Aboriginal and Torres Strait Islander · Environmental management

---

Justin McCaul is a descendent of the Mbarbarum people of far north Queensland. This paper derives from my Ph.D. thesis research analysing native title through the lens of deliberative democracy theory.

---

J. McCaul (✉)
Australian National University, Canberra, ACT, Australia
e-mail: justin.mccaul@anu.edu.au

 
N. Moodie and S. Maddison (eds.), *Public Policy and Indigenous Futures*, Indigenous-Settler Relations in Australia and the World 4,
https://doi.org/10.1007/978-981-19-9319-0_4

## 4.1 Introduction

The purpose of this collection of articles is to highlight Indigenous policy expertise in relation to a range of policy fields. It seeks to highlight what self-determination looks like when Aboriginal and Torres Strait Islander communities develop community-owned responses to complex policy challenges. In this chapter, I will argue that an example of an Indigenous led policy development and expertise in the area of land management, known as Caring for Country.[1]

In this chapter, I seek to make two points regarding Caring for Country as an example of Indigenous led policy expertise and success. The first is uncontroversial from the perspective of Aboriginal and Torres Strait Islander people. Because Aboriginal and Torres Strait Islander people never ceded sovereignty, we retain a right to self-government.[2] Caring for Country is, therefore, an exercise in self-government.

Since the *Mabo* decision, Aboriginal and Torres Strait Islander people have reclaimed large areas of their traditional lands and waters under native title legislation. Yet, within Australia, native title is limited to recognition of Aboriginal and Torres Strait Islander people's prior right in land. It is a proprietary right only. However, legal scholars such as Lisa Strelein (2009) and Mick Dodson (1995) argue that inherent to the recognition of native title is recognition of Aboriginal and Torres Strait Islander peoples' political right to self-government. As Strelein states,

> despite the compromise to the sovereignty of the Crown and the tendency of the courts to limit and contain it, native title is not just a property right nor merely an interest in the land. It is also a self-government right. It is a right or title in a group to order their own affairs, at least in relation to land. (2009, p. 123)

The second point is somewhat more novel. I seek to argue Caring for Country is an example of 'deliberative policymaking'. Deliberative democracy challenges the exclusive role of government and bureaucrats to decide policy by arguing that people affected by political decisions or public policies must be included in the decision-making process (Bächtiger et al., 2018; Elstub & McLaverty, 2014; Kahane et al., 2010). Advocates of deliberative democracy argue political decisions and policy made in this manner are more legitimate in the minds of citizens than decisions made exclusively by elites such as politicians and bureaucrats.

Deliberative policymaking challenges the traditional approach to public policy in which it is governments and parliaments, supported by bureaucrats and/or experts, to develop policy. In this traditional approach, the public, a segment of the public, the media, or advocates identify a problem requiring government intervention. The problem and any associated issues are analysed by politicians and bureaucrats in order

[1] Caring for Country also includes sea country. However, all of my experience with Caring for Country groups was land based so I will not discuss managing sea country.

[2] Various terms—self-determination, self-government, sovereignty—can be used to articulate Indigenous people's desire for political autonomy, our resistance to ongoing settler colonialism, and demands that a new relationship between Indigenous people and the state be forged through negotiation. For the sake of simplicity and in keeping with the aim of this book to demonstrate Indigenous policy expertise, I use the term self-government.

to produce a set of well-considered policies and strategies in response. Governments and bureaucrats decide upon the most suitable option and then set about implementing the policies to address the original problem. The effectiveness of the policy and its outcomes are then monitored (hopefully) and adjustments are made as needed (Bridgman & Davis, 1998, pp. 24–26; Maddison & Denniss, 2009).

Under the traditional policy approach of ‘government’, citizens are expected to passively acquiesce to government policies. By contrast, under a ‘governance’ approach, citizens take an active role in developing solutions to societal problems (Warren, 2009). Understood in this light, Caring for Country is an example of deliberative policymaking because it seeks to shift policymaking from what Hajer and Wagenaar (2003, p. 1) calls ‘government to governance’. Through Caring for Country, Aboriginal and Torres Strait Islander people challenge the state’s usual top down role in policymaking by taking an active role in the governance of Country.

Under the traditional government approach, Aboriginal and Torres Strait Islander people are framed as a problem to be solved or a gap to be closed, with the Australian settler state unquestionably accepted as the solution to Indigenous problems rather than the problem itself (Nakata & Maddison, 2019, p. 412). The ongoing failure of government-administered Indigenous affairs policy, which has relied on this traditional approach, is well documented (Moran, 2016; Sanders et al., 2002; Sullivan, 2011, 2013).

We often think of Indigenous peoples’ claims to self-determination and self-government as taking place within traditional institutions of the settler colonial legal and political system such as courts, parliaments, or even the public sphere. Day-to-day encounters between Aboriginal and Torres Strait Islander people and the state in policy and planning spaces can be easily overlooked. However, as settler colonialism scholar Elizabeth Strakosch (2019) explains, despite the settler state’s efforts to make Indigenous policy a neutral and benign space, domestic policy is a key site of political encounter between Indigenous peoples and the settler state:

> I argue that domestic policy, despite its focus on administration and technical ‘best practice’, is the key space where the Australian state encounters Aboriginal and Torres Strait Islander polities and seeks to resolve colonial conflict in its favour. (Strakosch, 2019, p. 115)

As Strakosch highlights, through public policy the state seeks to maintain is dominance over Aboriginal and Torres Strait Islander people and their lands. Nevertheless, as I will argue, the encounters between Aboriginal and Torres Strait Islander people through Caring for Country, are important sites of contestation and challenging state-centred public policy. These are sites of political relations between Indigenous authority and the state. The exercise of self-government manifests through deliberation on public policy.

In this Chapter I draw on my professional experience working with Aboriginal communities and Indigenous ranger groups on Caring for Country projects between 2005 and 2014. During this period, I worked with two conservation, non-government organisations (NGOs): the Australian Conservation Foundation (ACF), with a focus on environmental policy and advocacy; and Bush Heritage Australia, with a focus on practical land management. In both roles, my job was to broker relationships with Aboriginal communities[3] and organisations undertaking or seeking to undertake land management projects on Country. All of this work was across northern Australia—a vast area stretching from Cape York Peninsula in Queensland, the Gulf of Carpentaria and Arnhem Land in the Northern Territory, across to the far northern Kimberley region of Western Australia. I had what could be described as a 'birds eye view' of engagement between Aboriginal groups and government bureaucrats in a sometimes combative and messy cross cultural negotiating space mixing environmental management and native title rights. It was clear to me that Caring for Country was more than a process for land-use decisions. It was a process for the group to assert their right to be self-governing.

Brokering relationships was quite organic. A potential partnership could begin with a representative of the community or organisation interested in discussing their land management aspirations contacting either a colleague or myself. Alternatively, I would hear of a certain group doing good work or I would meet someone at a native title conference or land and sea forum and begin a discussion of what might be possible through a collaboration. Many but not all had claimed native title or were in the process of doing so. Initial discussions could take weeks or even months. If an agreement was reached, I was able in most situations to offer seed funding to get things moving. Or, if the Aboriginal group had already established a Caring for Country program, simply invest in their work. Funding was usually modest, so securing other partners and additional funding was always important.

Key to securing other funding was to produce a strategic document such as a draft plan of management, developed through a participatory planning process with the community. It would include a list of known environmental threats, sites of natural and cultural significance, and important indicators of environmental and cultural health. Most importantly, it captured the vision and aspirations of the community. Once documented, we would try to attract interest from either government, other NGOs, a philanthropic organisation, or all three. Outcomes were mixed as some groups found their feet—and other support and funding—quickly, while others found they had internal governance issues to resolve first.

In summary, this chapter aims to demonstrate that Caring for Country is an example of Indigenous led policy success because it is not simply a space for land-use planning and environmental management policy, but also as a forum for Aboriginal and Torres Strait Islander people to exercise self-government.

[3] My professional experience in native title and Caring for Country has predominantly been with Aboriginal groups in northern Australia. Accordingly, I limit my references to 'Aboriginal people and groups' rather than 'Aboriginal and Torres Strait Islander' people. I also use the term 'Indigenous' in this paper.

## 4.2 *Mabo* and Native Title as the Right to Self-government

Before discussing Caring for Country as an example of Indigenous led public policy success, I will discuss the importance of the *Mabo* decision and native title in the emergence of Caring for Country as distinct area of Indigenous led public policy.

It is well known the High Court of Australia in the historic *Mabo* decision (Mabo v State of Queensland (No. 2) HCA 23, 1992) declared rights of the Meriam people of the Torres Strait Islands had survived British annexation and 'that the Meriam people are entitled as against the whole world to possession, occupation, use and enjoyment of the lands of the Murray Islands'. Further, the Court found that where connection to traditional lands could be proved, the common law would now recognise and protect this right in places where native title had not been extinguished by the grant of an interest inconsistent with native title such as to another party, by legislation, or an action of the executive.

*Mabo* was deemed a 'judicial revolution' (see Stephenson & Ratnapala, 1993) and widely expected to lead to greater economic and political empowerment of Aboriginal and Torres Strait Islander people. This, however, has not been the case, with many groups struggling to convert their native title rights into substantive political power or sustainable economic benefit for their communities. As a result, many Aboriginal and Torres Strait Islander people hold cynical views of the benefits of native title (Brennan et al., 2015).

While native title may not have delivered on the promise of *Mabo*, the decision nonetheless confirmed Australia now has two systems of law in relation to land (Jackson et al., 2017, p. 177). And despite its limitations, native title has forced the reconsideration of established assumptions about who owns land in Australia, who gets to decide what it is used for and how, and who is responsible for its management, all of which has necessitated a transformation in land related public policies.

Native title, along with statutory land rights regimes and other legislative mechanisms, has made possible an expansive 'Indigenous estate'—land held and managed by Aboriginal and Torres Strait Islander people of various tenure types (Altman et al., 2007, p. 5). The recognition of native title expanded opportunities for Indigenous led land management across large areas of Australia, with Aboriginal and Torres Strait Islander people owning and/or managing more than 20 percent of Australia's landmass (Jackson et al., 2017, p. 177). Caring for Country has become a vehicle for asserting rights in land, re-establishing cultural institutions and decision-making, and deriving a living from the modern political economy (Altman & Kerins, 2012).

A complete discussion of native title case law and its effect on different forms of land tenure is not possible, but after the *Mabo* decision, two other cases would clarify where native title could co-exist with other titles (e.g. Wik Peoples v The State of Queensland & Ors; The Thayorre People v The State of Queensland & Ors [1996] HCA 40, 1996) and where it is susceptible to extinguishment (e.g. Western Australia v Ward [2002] HCA 28, 2002).

In *Wik,* the High Court determined that pastoral leases did not confer exclusive possession and native title could co-exist with pastoral leases to the extent of any inconsistency. Despite the uproar the *Wik* decision provoked, the Court made clear that native title was a lesser right than the rights held by pastoralists. In the event of any clash, the rights of native titleholders would have to yield 'if there was inconsistency between those rights and the rights asserted by the native title holders, the latter had to yield to that extent to the rights of the grantees' (Wik Peoples v The State of Queensland & Ors; The Thayorre People v The State of Queensland & Ors (1996) HCA 40, 1996).

In *Ward,* the High Court disappointingly overturned the earlier decision of the Full Federal Court (Western Australia v Ward and Ors (2000) 170 ALR 159, 2000), that determined reserves and national parks did not necessarily extinguish native title. The High Court instead declared that the reservation of lands for nature reserves and national parks,

> was inconsistent with any continued exercise of power by native title holders to decide how the land could or could not be used, but not necessarily inconsistent with the right to continue to use the land according to traditional laws and customs.

Despite the vulnerability of native title to extinguishment in areas such as reserves and national parks, these areas often include sites of spiritual and cultural significance to Aboriginal people. This understanding prompted governments to develop policies to deal with Indigenous demands for land justice as well as strategies and programs to 'reconcile the competing imperatives of ecosystem protection and Indigenous rights and cultural heritage' (Bauman et al., 2013, p. 10). As Jackson comments, 'common law recognition of native title and social justice initiatives has created a nexus between environmental and natural resource management processes inclusive of Indigenous values and interests and the participation of Indigenous communities' (Jackson, 2006, p. 20).

## 4.3 Native Title as Self-government

We know that before *Mabo,* the prevailing assumption was that Aboriginal and Torres Strait Islander people possessed no rights in land that the Crown could recognise. This non-recognition was based on the concept of *terra nullius* and the belief that Aboriginal people possessed no 'settled law' the British could formally recognise (Strelein, 2009, p. 2). The *Mabo* decision overturned this erroneous presumption. Mick Dodson remarked shortly after the enactment of the *Native Title Act 1993,*

> If native title is a title based on our laws and customs, it is an absurd position if our title to land is recognised but the laws and customs which give meaning to that title are treated as if they do not exist. The Australian legal system must take the further step of accepting that native title is inseparable from the culture which gives it its meaning. (Dodson, 1995, p. 2)

Likewise legal scholars such as Lisa Strelein and Jeremey Webber have argued native title is more than a simple proprietary right, it is a right to self-government given it recognises 'Aboriginal societies as societies' (Webber, 1999, p. 601). As Strelein states,

> Despite the compromise to the sovereignty of the Crown and the tendency of the courts to limit and contain it, native title is not just a property right nor merely an interest in the land. It is also a self-government right. It is a right or title in a group to order their own affairs, at least in relation to land (Strelein, 2009, p. 123).

Similarly, as Webber explains,

> Claims of indigenous self-government are often treated as though they are quite separate from claims of native title. .... [I]ndigenous title necessarily involves that one recognises the presence of autonomous legal orders that have their origin outside the settler society's legal system. Once one acknowledges (as the High Court does) that indigenous orders are dynamic entities, with the capacity to develop and change, the governmental dimension become clear. It is no exaggeration to say that the recognition of indigenous title necessarily involves the recognition of self-government (Jeremy Webber in Ivison et al., 2000, p. 87).

As Aguis et al. (2007, p. 197) acknowledge in their study of native title negotiations in South Australia, while native title has not delivered Aboriginal and Torres Strait Islander people exclusive possession over their traditional domains, it has posed deep questions over the primacy of settler claims to place in Australia. Agius et al. explain that planning under native title is a space for contesting the 'politics of place',

> Nevertheless, the spatial politics of Native title through agreement making presents some hope for constructing new geographies and changing the place that Australia is. It has provided a vehicle for Indigenous groups to participate more directly, and at times with greater authority, in the politics of place (Agius et al., 2007, p. 196).

They go on to state,

> What has evolved is a complex political and social process that establishes spaces for dialogue and exchange between participating groups to facilitate negotiation of Native title in fair and just ways. Participation is voluntary, and all parties enter the negotiation space as equals. Building real working relationships and mutual understanding between the parties is supported. The focus is not centred on where Native title does or doesn't exist, but on the practicalities of how coexistence might be realised (Agius et al., 2007, p. 197).

For Agius et al., native title has a strong self-government dimension for relations with the state, and for place-based policy and planning.

To illustrate my point further that native title is self-government, I describe a scenario I observed many times when working with Aboriginal groups on Caring for Country projects in meetings with government representatives, usually from the departments of environment, parks and wildlife, or Indigenous affairs. The purpose of the meetings was to discuss a possible Caring for Country project. They would begin with a discussion of land management issues—weeds, feral animals, the prevalence of wildfires—as well as of resources necessary for establishing a Caring for Country program such as funding for a facilitator, employing a team of Indigenous rangers, office space, equipment, vehicles etc. Soon, though, members of the native title group would begin to discuss other community issues such as the need for better housing, more teachers in the school, more services in the health centre, or reforms to policing policies. At this point the government representatives would look a bit confused before stating, 'We are here only to discuss native title issues'. To which the group would respond, 'We are talking about native title!'.

What this anecdote demonstrates is that both native title and Caring for Country are much more than a means for managing land. For Aboriginal and Torres Strait Islander people both are more holistic, incorporating a right to make decisions in relation to both land *and* the people that live on that land. In short, incorporating self-government.

Another example of this is the Neale et al. (2019) case study of fire management on Dja Dja Wurrung Country. Neale et al. question whether we might describe environmental management collaborations between the Victorian government and Traditional Owners as 'experiments in decolonising':

> So, are Dja Dja Wurrung peoples' engagements with bushfire decolonising, in this sense? While, in Australia's southeast, it is unlikely that any Aboriginal peoples will soon gain full autonomy over their country, we suggest that bushfire management is a site of emerging experiments in the redistribution of legal and political authority over country. Thinking broadly, such collaborative projects in environmental management are experiments in the sense that an experiment is 'a device to materialise questions' or, as Fortun states, something that allows for 'the emergence of questions that could not be asked before'; it opens 'the possibility of envisaging ways in which knowledge/space/society could be transformed' (Neale et al., 2019, p. 346).

Though Neale et al. employ the term 'decolonising' rather than self-government, the intention is the same: to situate Caring for Country as a space where questions of self-government emerge through the assertion of Aboriginal authority over traditional lands and waters.

The right to self-government as an extension of recognition of native title is a point obvious to Aboriginal and Torres Strait Islander people. Yet Australia's continued denial to recognise the inherent political right to self-government in native title, is at odds with other settler colonial states such as Canada (Perry, 2014; Russell, 2006, pp. 239–240). As Goenpul woman and Aboriginal academic Aileen Moreton-Robinson explains, Australia treats Aboriginal and Torres Strait Islander people as trespassers on our own lands until we can prove native title (Moreton-Robinson, 2015, p. 69).

Following Neale et al. and Agius et al., the engagement between Aboriginal and Torres Strait Islander people and the state through Caring for Country, is a policy space that surfaces questions of self-government and Traditional Owner rights to make decisions in relation to land and people. It is in this (re)framing of the meaning of native title as self-government, or as an exercise in decolonisation, that is, I argue, at the heart of Caring for Country as successful Indigenous led policy.

## 4.4 Caring for Country: A Policy Context Overview

In this section, I provide an overview of the development of Caring for Country as a distinct area of Indigenous led public policy. Firstly, Caring for Country can be understood as the relationship Aboriginal and Torres Strait Islander people have with their traditional lands and waters. It is the strong cultural, spiritual, and physical attachment, and sense of belonging that Aboriginal people feel to a specific place, expressed as 'my country' (Rose, 1996; Weir et al., 2011). To Care for Country is thought of as a reciprocal relationship, often expressed by Aboriginal and Torres Strait Islander people in the refrain: 'If we look after country, country will look after us'.

Secondly, it can also be understood as a cross-cultural collaborative space between Indigenous and non-Indigenous groups necessitating a 'two way' approach (Emile Ens in Altman & Kerins, 2012, p. 45) to managing the natural environment that utilises both Indigenous traditional knowledge and Western scientific knowledge (Hoffmann et al., 2012). Activities include biodiversity conservation, traditional fire management, water resource management, feral animal and weed control, environmental monitoring, threatened species protection, and even aspects of border control and monitoring of illegal fishing.

Thirdly, and the focus of this chapter, it has come to be understood as a unique policymaking space that blends native title, Indigenous ecological knowledge, and the right to self-government, to manage the environmental, cultural heritage, and natural resources of Country. As a result of native title and land rights legislation, Caring for Country activities are undertaken across a variety of tenure types including pastoral lands, state owned national parks, co-managed national parks, and on Indigenous owned lands such as Indigenous Protected Areas (IPAs).

In public policy terms, Caring for Country was first used in a 1991 report by the Australian National Parks and Wildlife Service with funding from the then Aboriginal and Torres Strait Islander Commission (ATSIC) (Young et al., 1991). The report highlighted a lack of knowledge about land degradation on Indigenous held lands, and recognised that Aboriginal land managers faced challenges in accessing mainstream government funds to tackle environmental issues. It identified a need to provide recurrent funding to Indigenous groups to enable them to plan and organise their land management activities, support greater Indigenous participation in decision-making on grants and funding, and generally advocated for raising the profile on Indigenous land management needs and opportunities. The report argued land management activities of Aboriginal people were important and legitimate, yet grossly underfunded and not well supported by government (Young et al., 1991). Shortly after this report was published, the Northern Land Council (NLC) established a Caring for Country Unit in 1995. As Storrs and Cooke explain,

> The Caring for Country Unit's (CFCU) catchcry is 'the land needs its people'. That is to say it operates under a philosophy of building the capacity of the people to look after their land (which is empowering) versus the philosophy of setting up an agency to look after the land on behalf of the people (which is disempowering) (Storrs & Cooke, 2001).

Caring for Country, however, does not equate seamlessly with the Western scientific concept of environmentalism. Several high-profile disputes between conservationists and Indigenous groups over the issue of 'sustainability' occurred during my time at ACF that illustrate this point. One was the volatile debate over wild rivers legislation in Cape York Peninsula (Neale, 2017, p. 30) and the other a proposal to build a large natural gas refinery at James Price Point near Broome in Western Australia (Burnside, 2013). Both cases involved development proposals that promised significant economic benefits to local and regional Aboriginal communities, but had large environmental and social impact concerns for local communities.

The tension in both disputes was a perception that environmental concerns were a higher priority than the rights of Aboriginal people to make decisions in relation to their traditional lands. In relation to the Wild Rivers legislation in Queensland, Noel Pearson, a Guugu Yimithirr Traditional Owner and lawyer claimed conservation was 'a new wave of colonialism' (Owens & Wilson, 2010), denying Aboriginal people the right to economic development and self-determination. These and other cases attest to a potential incommensurability between 'green' and 'black' notions of sustainability (Vincent & Neale, 2017).

The seeds of Caring for Country can be traced to events before *Mabo* including the 'homelands movement' of the 1970s and the enactment of statutory land rights regimes such as the *Aboriginal Land Rights (Northern Territory) Act 1976* and the adoption of self-determination as the basis of Indigenous policy at the national level in 1972 (Altman & Whitehead, 2003, p. 2). The homelands movement saw Aboriginal people return 'home' to re-occupy their traditional lands and to also escape the many social problems and experiences of discrimination that Aboriginal people suffered living in major towns and cities (Glaskin, 2007, p. 202).

Joint management or co-management of state-owned national parks was also important for creating opportunities for Indigenous led environmental management and strengthened calls by Aboriginal groups for resources to undertake land management activities on Country. As Bauman et al. explain:

> Seen through such a lens, co-management is a distinctive and complex single form of governance in which traditional owners, Indigenous organisations and bureaucracies interact to produce shared outcomes. Achieving shared outcomes requires the meeting of 'top-down' bureaucratic policy and decision-making and 'bottom-up' decision-making, needs and interests. (Bauman et al., 2013, p. 11)

Despite these positive changes, Australian governments and policy makers were slow to grasp the potential of Indigenous community based natural resource management (Altman & Whitehead, 2003, p. 2). Then in the late 1980s and the early 1990s, with global attention on the issues of sustainability and environmental degradation, two events would be influential to the future growth of Caring for Country.

The first was the release of the 'Our Common Future', or Brundtland, Report (Brundtland, 1987), which highlighted the need for governments and local communities to do more to tackle the increasing degradation of the natural environment. The second was the 1992 Earth Summit in Brazil and subsequent Rio Declaration calling for greater recognition and involvement of the world's Indigenous people in tackling these issues. Recognition of Indigenous rights was now coupled with calls to utilise Indigenous traditional knowledge in strategies to combat environmental degradation globally and domestically. Principle 22 of the declaration states:

> Indigenous people and their communities and other local communities have a vital role in environmental management and development because of their knowledge and traditional practices. States should recognize and duly support their identity, culture and interests and enable their effective participation in the achievement of sustainable development (Antrim, 2019, p. 189).

In Australia, governments and conservation groups responded by developing strategies and policies to recognise Indigenous people's rights in land in relation to environmental management (Hill et al., 2013, p. 6). By now, decentralised decision-making in relation to environmental management in Australia was evolving through initiatives such as Landcare that sought to give local landholders such as farmers greater autonomy and decision making responsibilities rather than maintain a 'top down' decision-making model (Taylor et al., 2010). Then, in 1997, the Natural Heritage Trust (NHT) expanded upon the Landcare approach, making funds available to regional natural resource management (NRM) bodies, non-government organisations (NGOs), and landowners' to design and deliver NRM projects to suit local circumstances (Robinson and Lane in Walker et al., 2013, p. 397).

The objectives of both Landcare and NHT were to empower local level parties most affected by environmental issues to make decisions and take responsibility (Environment and Communications References Committee, The Senate 2015); an ideal that aligned with Aboriginal people's desire to have decision-making power over Country vested with them, instead of bureaucrats sitting in distant government offices.

In 1997, the first Indigenous Protected Areas (IPA) were established after discussions between the Federal government and Aboriginal groups about the inclusion of Indigenous held lands into the national reserve system (Smyth, 2006). IPAs are an internationally recognised form of protected area that are on Indigenous-held lands. Their declaration is a voluntary contribution to the protection of Australia's unique environmental and cultural heritage, and account for more than 48 million hectares of land and around 36 per cent of Australia's total system of protected areas (Stoeckl et al., 2016).

Then, in 2007, the Federal Government announced Working on Country,[4] including $47.6 m funding over four years for Indigenous rangers (Hill et al., 2013, p. 80; See Mackie & Meacheam, 2016, p. 160). Since 2007, both Caring for Country activities and the number of IPA declarations has expanded significantly, with more groups undertaking community-based planning and establishing ranger groups on their own lands (Altman & Kerins, 2012, p. 14).

The number of reports highlighting a range of benefits, opportunities, and outcomes in relation to Caring for Country has also expanded significantly (Altman & Jackson, 2014; Altman & Kerins, 2012; Altman et al., 2007; Hill et al., 2013; Mackie & Meacheam, 2016; Weir et al., 2011; Woodward et al., 2020). Studies have

---

[4] The term Working on Country is the government name for the funding of Caring for Country activities. Between 2008–2018 the program was renamed Caring for Our Country, blurring the lines between what is a distinctly Indigenous mode of environmental management and government funding.

highlighted benefits such as better health, better education, greater workforce participation, creating Indigenous businesses, and lower expenditures on public health, policing, welfare (Campbell, 2015; Schultz & Cairney, 2017). Others have emphasised economic benefits and social well-being (Bueren et al., 2015; Gorman & Vemuri, 2012; Jarvis et al., 2018).

Additional studies have argued that increased engagement with Aboriginal communities through Caring for Country leads to better knowledge and understanding for landscape protection and environmental restoration, while also supporting Indigenous communities to fulfil cultural obligation to look after Country (Roughley & Williams, 2007; Woodward et al. 2020). Yet challenges remain such as inadequate funding (Williams and Hill in Taylor et al., 2010) and the lack of genuine inclusion of Indigenous knowledge in environmental management (Samuel, 2020; See also Hill et al., 2013). The Australian government recently signalled its strong support for Caring for Country by announcing an additional $700 m in funding for Indigenous rangers undertaking Caring for Country work beginning in 2021 and extending to 2028 (Wyatt & Ley, 2020).

A report entitled 'Strong on Country' states the past 40 years have been a 'quiet revolution' in the growth of Indigenous land and sea management as Aboriginal and Torres Strait Islander people reclaim their role as managers of their traditional lands and waters (Putnis et al., 2021). There are now some 129 Indigenous ranger groups (Putnis et al., 2021, p. 13), and since 2007, Caring for Country projects have created more than 2100 full-time, part-time, and casual jobs in land and sea management around the country, providing meaningful employment, training and career pathways for Aboriginal and Torres Strait Islander Australians (Department of Prime Minister & Cabinet, 2016).

Beginning with the legislative recognition of Indigenous land rights in the 1970s and bolstered by international trends to incorporate Indigenous knowledge into environmental protection as well as the subsequent creation of IPAs, Indigenous led decision-making has become a defining feature of the Caring for Country policy context, and part of the environmental policy 'lexicon' in Australia (Pleshet, 2018, p. 184). I now turn to discuss Caring for Country as a form of deliberative policymaking.

## 4.5 Caring for Country as an Example of Deliberative Policymaking

In this section, I seek to illustrate how Caring for Country can be considered an example of 'deliberative policymaking' or 'democratisation through policy discourse' (Torgerson in Hajer & Wagenaar, 2003, p. 113) and what Mark Warren terms 'governance driven democratization' (Warren, 2009). Democratization entails a deepening role of citizens in self governance and decision making. To do this I draw on two cases, one involving the Wunambal Gaambera Healthy Country

Plan process and the second a collaborative Indigenous fire management process in Victoria (Neale et al., 2019). I also draw on the work of Agius et al., and native title negotiations in South Australia (Agius et al., 2007), to illustrate how native title has transformed the 'politics of place' to enable greater involvement of Indigenous people in environmental management through Caring for Country projects.

As an extreme minority in Australia's model of representative democracy, our political voice is underrepresented in formal decision-making forums such as parliaments. As such, deliberative democracy holds an obvious appeal for Aboriginal and Torres Strait Islander people with its emphasis on active citizen participation and a talk centric approach to decision-making.

As Elstub et al. explain, the normative claim which unites delibertaive democracy as a body of thought is that political decision-making should be talk-centric rather than vote-centric. 'Rather than merely constituting the aggregation of individual preferences, collective decisions should emerge from public reasoned discussion and debate' (Elstub & McLaverty, 2014, p. 1). Further, it stresses that people most likely to be affected by a decision should have the right, capacity, and opportunity to participate in the process of making of those decisions (Dryzek, 2001). Deliberation becomes a process of mutual justification, where participants offer reasons for their positions, listen to other views, and reconsider their preferences in the light of new information and argument (Gutmann & Thompson, 2004, p. 3).

As Iris Marion Young (2000, p. 17) explains, traditional democracy tends to reinforce political, social, and economic inequalities that preserve the privilege of the powerful. Deliberative democracy espouses the principles of inclusion, equality, reasonableness and accountability in decision-making as a means of counteracting this trend. In deliberative democracy, citizens are central to the process of developing proposals on how best to solve a policy problem. Participants use open discussion, reasoning, persuasion, and normative appeals to convince others to support their proposal (Young, 2000, p. 22).

Politics often entails strategic bargaining—'give me this and I'll give you that'—with each party concerned with maximising benefits and minimising costs. As Gutmann and Thompson explain, deliberative democracy asks both citizens and political representatives to exchange reasons for their decisions and be prepared to respond with an open mind to the reasons given by the other (Gutmann & Thompson, 2004, p. 3).

> The moral basis for this reason giving process is common to many conceptions of democracy. Persons should be treated not merely as objects of legislation, as passive subjects to be ruled, but as autonomous agents who take part in the governance of their own society directly, or through their representatives. (ibid)

The idea is that the process makes it less likely that purely self-interested or ill-conceived proposals survive scrutiny. Advocates of deliberative democracy view it as useful in situations affected by polarising points of view, power imbalances, multi-stakeholder interests, and for bridging cultural differences.

By contrast, an aggregative approach to democracy is the expression of personal preferences through voting as the primary political act. Deliberative democracy seeks to increase 'participatory democracy' (Uhr, 1998, p. 11) so that it engages marginalised groups in the process of developing public policy (Heller & Rao, 2015, p. 4).

Also, deliberative democracy asks parties not to bargain from fixed positions but rather remain open minded to the possibility of being convinced by 'unforced force of the better argument' (Steiner, 2012, p. 4). Advocates of deliberative democracy claim it generates decisions that are more legitimate because they have been produced through public participation, citizens' commitment and cooperation in the decision-making process. They claim the process works to foster mutual understanding between individuals, avoids exclusion, and promotes better quality of decisions through discussion (Chambers, 2003, p. 317). In summary, deliberative democracy offers a process for facilitating broader perspectives on issues of public policy (Gutmann & Thompson, 2004, p. 11) and 'broadening and deepening' democracy (Ercan & Dryzek, 2015, p. 242).

There are numerous studies of deliberative democracy in relation to environmental management including community planning for the protection and management of freshwater resources in New Zealand (Arunachalam et al., 2016), river basin governance in Brazil (Abers & Keck, 2006), logging of old growth forests in British Columbia (Torgerson in Hajer & Wagenaar, 2003, pp. 113–138), and the 2012 Tasmanian Forest Agreement (TFA)(Schirmer et al., 2016).

In relation to Indigenous peoples, it has been applied to explain traditional systems of governance and decision-making (Reedy et al., 2020), and to address historical injustice through the process of reconciliation (Ivison in Kahane et al., 2010, p. 115). As Hill and Williams explain (in Taylor et al., 2010), there can be a significant difference between Indigenous and non-Indigenous understandings of the world, so deliberative democratic processes may be more capable of facilitating meaningful participation between these two groups, particularly in the context of land management.

More recently, in line with the aims of this chapter, deliberative democracy theory has been used to discuss environmental policymaking in Australia between Indigenous peoples and the state (Davis, 2021). Like Davis, I suggest that relations between the state and Aboriginal and Torres Strait Islander people in environmental policymaking provide insight into how politically marginalised groups such as Aboriginal and Torres Strait Islander people engage in democracy-making through policy discourse:

> How, marginalised voices are recognised in the public sphere and connected to the state to influence policy outcomes is important. Transmission of Indigenous environmental messages is potentially more effective when it happens between Indigenous empowered space and state empowered space. Indigenous deliberative forums connected to the state which enable the sharing of reasons with effective transmission to state policy making is a democratic innovation which may make this possible (Davis, 2021, p. 384).

What Davis terms 'Indigenous empowered space' I understand to be Country. The meeting of state officials 'on Country' with Aboriginal and Torres Strait Islander people is one of the reasons for its success. Having outlined the fundamental principles of deliberative democracy I now discuss why Caring for Country is an example of deliberative policymaking.

During my time at Bush Heritage Australia, I worked with the Wunambal Gaambera Traditional Owners of the north Kimberley region. They had developed a Healthy Country Plan that articulated their aspiration to look after Country. As the Wunumbal Gaaambera Traditional Owners explain in their plan, Caring for Country is done in accordance with,

> our Wanjina Wunggurr Law... protecting and sharing our cultural places as our traditional Law says. We will be respected as the proper owners and managers of Wunambal Gaambera Country and our future generations, will have the cultural knowledge of our elders, our country will be giving us and our future generations a healthy life (Wunambal Gaambera Healthy Country Plan, 2010).

For Wunumbal Gaambera, to manage Country is to follow Wanjina Wunggurr Law. That is to say, only Wunumbal Gaambera law and political authority can legitimately manage Country. As Moorcroft et al., (2012) explains, healthy country plans helps Indigenous Traditional Owners to communicate more strategically with external stakeholders such as governments. Planning becomes a structured negotiation rather than an externally imposed process. With their own ideas and aspirations incorporated into their plan in their own language, Wunambul Gammbera Traditional Owners can better control the planning process, creating an more equitable conservation planning space.

Likewise, as Neal et al. explain, their experience of co-operative bushfire management in Victoria, helps to highlight the existence of persistent inequalities in conservation planning and management. Legal recognition (native title) to manage Country is used to highlight social, economic, and political inequality and marginalisation. As an act of decolonisation, collaborative bushfire management becomes an experiement in the reistribution of legal and politcal authority over Country. Indigenous led environmental management creates room for forms of Indigenous knowledge so that the planning process may be transformed. So, to undertake traditional burning (*djandak wi* or 'healthy fire') on Dja Dja Wurrung Country in Victoria, for ecological and cultural reasons the areas to be treated with fire must be decided by Dja Dja Wurrung Elders before choosing select Dja Dja Wurrung people to work in partnership with settler bushfire agencies. By 'walking together', Indigenous and non-Indigenus parties needed to be open and vulnerable to change and incorporating the suggestions of others, particularly Dja Dja Wurrung Traditional Owners. Both Wunambal Gaambera and Dja Dja Wurrung are examples of Aboriginal groups integrating their political authority over land within environmental planning and policy processes. In the various Caring for Country meetings I participated, Aboriginal groups would insist government and industry representatives to listen to and respond to Indigenous concerns, challenging their assumed role as primary decision maker. Caring for Country becomes a forum for reflexive policy discourse in which the state's expertise (or bias) is challenged by citizen expertise (local knowledge) (Torgerson, in Hajer and Wagenaar, 2003, p.136).

As Hill et al. argue Aboriginal people driving the planning process is important. This is to ensure non-Indigenous conservation planners and government representatives listen to and take on board Indigenous governance structures and local protocols, priorities, and traditional ecological knowledge (Hill et al., 2013, p. 51). Strategic documents such as Healthy Country Plans developed through 'Country-based planning' are an expression of Indigenous authority to manage Country. Country-based planning is the process by which Aboriginal and Torres Strait Islander peoples identify their management aspirations and strategies across the whole of their traditional territories. It seeks to join people's responsibilities to manage Country with the authority to meet these responsibilities (Hill et al., 2013, p. 51).

Perhaps a more powerful articulation of Caring for Country as a political authority and cultural responsibility and not simply a land use activity, is expressed by Langton who explains that nature of Aboriginal proprietary interests are a 'spiritual bequest' from one generation to another:

> Property relations are thus subservient to the fundamental social relationship with the sacred ancestral past from which the legacy is derived and which bestow not just property rights, but a social world, and for particular people, arduous religious duties and responsibilities which are transmissible across generations (Marcia Langton in Godden & Tehan, 2010, p. 76).

As Agius et al., explain in their research on native title negotiations in South Australia, despite its limitations, native title has created space for a reconsideration of politcal relations between Indigenous people and the state. It has created a space for dialogue on what coexistence may look like and recognition of Aboriginal jurisdiction through a 'politics of place'. Though these examples do not explicitly employ the language of deliberative democracy, the principles, values, and processes they describe are recognisably *deliberative* in that they advocate for exchanges between Aboriginal groups and the state to be approached not as a contest between adversaries, but a process for parties to develop mutually acceptable negotiation protocols and processes:

> Such considerations by the parties have resulted in processes that address power imbalances, capacity needs, understandings, uncertainties, relationships, respect and trust, and the defining of responsibilities to enable the making of informed decisions. Time has also been an important element of getting the process right, and the Main Table parties all endeavour to ensure sufficient time is available to address all participants' needs—something parties to litigated native title proceedings rarely have. (Agius et al., 2007, p. 198)

As Mark Warren explains, the rationale behind 'democratization through governance' is that policy and policy-making, usually led by bureaucrats, becomes a new front in democratization. Public distrust of political institutions and policymakers and greater policy expertise amongst citizens means democracy and democratization moves into the domain of policy and administration. The ability to influence policy relies less on elections (the traditional form of citizen participation in represenative democracy) and more on a greater role for citizens in policy development, planning, and administration. However, unlike Warren's assertion that greater democratization is 'elite driven' I argue the success of Caring for Country has been driven by Aboriginal and Torres Strait Islander people. It is a community based response to colonization and dispossession in which managing country is both an exercise of Indigenous jurisdiction and tackling important envionmental issues. If, as I have argued, Caring for Country is an example of deliberative policymaking, what is the significance of this claim? What makes engagement between Aboriginal groups and the state (and others such as non-government organisations) in the Caring for Country policy space, so unique? As Torgerson explains, in deliberative policymaking there is '…(t)he prospect that democracy might be enhanced not against or in spite of policy discourse, but through it' (Torgerson in Hajer & Wagenaar, 2003, pp. 113). Further, deliberative approaches to policymaking are an effort in 'expansive democracy':

> Expansive democracy is characterised by increased participation, either by means of small-scale direct democracy or through strong linkages between citizens and broad-scale institutions, by pushing democracy beyond traditional political sphere, and by relating decision-making to the person who are affected (Togersen in Hajer & Wagenaar, 2003, p. 3).

The preference for a more deliberative approach to policymaking by Aboriginal and Torres Strait Islander people can be viewed as democracy through policy discourse. It accentuates the political character of policy (Torgerson in Hajer and Wagenaar, 2003, p. 117). With Australia unable and unwilling to heed Indigenous demands for self-government, Caring for Country gives Aboriginal and Torres Strait Islander people and the state a way to talk to one another. As public policy, Caring for Country opens up a deliberative space for greater engagement with the state and Indigenous governance (self-government) as a way of overcoming the limits of the existing structures of representative democracy.

## 4.6 Conclusion

I conclude this chapter by reiterating that Caring for Country is an important area of Indigenous led public policy expertise. By combining native title, traditional ecological knowledge, and acts of self-government, Caring for Country opens up the opportunity for Aboriginal and Torres Strait Islander people to exercise self-government and engage in democratisation through policy discourse.

Caring for Country is a successful example of Indigenous led policy expertise because of its popularity and growth amongst Aboriginal and Torres Strait Islander people as a means to exercise Indigenous political authority over Country. This is despite the ongoing dominance of settler institutions and processes in policymaking and control of funding to broader Indigenous affairs policy. Nonetheless, Caring for Country can be a means of understanding how democracy can be deepened and improved through a deliberative, 'talk centric' approach to policymaking.

While this chapter seeks to emphasise positive features of Caring for Country it must be acknowledged that Aboriginal and Torres Strait Islander people remain limited in how much we can shape the policies that affect us to the degree we may like. Yet, despite the continued dominance of the state in shaping public policy,

Caring for Country has given Aboriginal and Torres Strait Islander people an opportunity to demonstrate what may be possible when the Australian state stops denying Aboriginal and Torres Strait Islander people aspirations to be self-governing.

## References


Abers, R. N., & Keck, M. E. (2006). Muddy waters: The political construction of deliberative River Basin governance in Brazil. *International Journal of Urban and Regional Research, 30*, 601–622.

Agius, P., Jenkin, T., Jarvis, S., Howitt, R., & Williams, R. (2007). (Re)asserting Indigenous rights and jurisdictions within a politics of place: Transformative nature of native title negotiations in South Australia. *Geographical Research, 45*, 194–202.

Altman, J., Buchanan, G., & Larsen, L. (2007). *Environmental significance of the indigenous estate: natural resource management as economic development in remote Australia.* Centre for Aboriginal Economic Policy and Research (CAEPR), ANU

Altman, J., & Jackson, S. (2014). Indigenous land and sea management 10.

Altman, J., & Kerins, S. (Eds.). (2012). *People on country: Vital landscapes, Indigenous futures.* The Federation Press.

Altman, J. C., & Whitehead, P. J. (2003). *Caring for country and sustainable Indigenous development: Opportunities, constraints and innovation* (CAEPR Working Paper 2014). CAEPR.

Antrim, L. N. (2019). The United Nations conference on environment and development. In A. E. Goodman (Ed.), *The diplomatic record 1992–1993* (pp. 189–210). Routledge.

Arunachalam, M., Singh-Ladhar, J., & McLachlan, A. (2016). Advancing environmental sustainability via deliberative democracy: Analysis of planning and policy processes for the protection of Lake Taupo. *Sustainability Accounting, Management and Policy Journal, 7*, 402–427.

Bächtiger, A., Dryzek, J. S., Mansbridge, J., & Warren, M. E. (Eds.). (2018). *The Oxford handbook of deliberative democracy.* Oxford University Press.

Bauman, T., Haynes, C., & Lauder, G. (2013). *Pathways to the co-management of protected areas and native title in Australia* (AIATSIS Research Discussion Paper). AIATSIS.

Brennan, S., Davis, M., Edgeworth, B., & Terrill, L. (Eds.). (2015). *Native title from Mabo to Akiba: A vehicle for change and empowerment?* Federation Press.

Bridgman, P., & Davis, G. (1998). *Australian policy handbook.* Allen & Unwin.

Brundtland, G. H. (1987). *UN World commission on environment and development* (Report of the World Commission on Environment and Development: Our Common Future). UN.

Bueren, M. van, Worland, T., Svanberg, A., & Lassen, J. (2015). Working for our country: A review of the economic and social benefits of Indigenous land and sea management. *Country Needs People.*

Burnside, S. (2013). James price point: Victory or loss? *Arena Magazine.*

Campbell, D. (2015). Aboriginal involvement in caring-for-country: An economic case study in primary preventative health. *Australasian Psychiatry, 23*, 623–625. https://doi.org/10.1177/1039856215612985

Chambers, S. (2003). Deliberative democratic theory. *Annual Review of Political Science, 6*, 307–326.

Davis, R. (2021). Australian Indigenous environment policy as a deliberative system. *Australian Journal of Political Science, 56*, 376–392. https://doi.org/10.1080/10361146.2021.1998342

Department of Prime Minister & Cabinet, (2016). Indigenous ranger programs [WWW Document]. https://www.niaa.gov.au/indigenous-affairs/environment/indigenous-ranger-programs

Dodson, M. (1995). From lore to law: Indigenous rights and Australian legal systems opinion. *Alternative Law Journal*, 2–2.

Dryzek, J. S. (2001). Legitimacy and economy in deliberative democracy. *Political Theory, 29*, 651–669.

Elstub, S., & McLaverty, P. (Eds.). (2014). *Deliberative democracy: issues and cases*. Edinburgh University Press.
Ercan, S. A., & Dryzek, J. S. (2015). The reach of deliberative democracy. *Policy Studies, 36*, 241–248.
Glaskin, K. (2007). Outstation incorporation as precursor to a prescribed body corporate. In K. Glaskin & J. F. Weiner (Eds.), *Customary land tenure and registration in Australia and Papua New Guinea, anthropological perspectives* (pp. 199–222). ANU Press.
Godden, L., Tehan, M. (Eds.), (2010). Comparative perspectives on communal lands and individual ownership: sustainable futures. Routledge.
Gorman, J., & Vemuri, S. (2012). Social implications of bridging the gap through 'caring for country' in remote Indigenous communities of the Northern Territory, Australia. *Rangel Journal, 34*, 63. https://doi.org/10.1071/RJ11037
Gutmann, A., & Thompson, D. (2004). *Why deliberative democracy? (1st edn)*. Princeton University Press.
Hajer, M. A., & Wagenaar, H. (2003). *Deliberative policy analysis: Understanding governance in the network society*. Cambridge University Press.
Heller, P., & Rao, V. (Eds.). (2015). *Deliberation and development: Rethinking the role of voice and collective action in unequal societies*. The World Bank.
Hill, R., Pert, P. L., Davies, J., Robinson, C. J., Walsh, F., & Falco-Mammone, F. (2013). Indigenous land management in Australia: Extent, scope, diversity, barriers and success factors.
Hoffmann, B. D., Roeger, S., Wise, P., Dermer, J., Yunupingu, B., Lacey, D., Yunupingu, D., Marika, B., Marika, M., & Panton, B. (2012). Achieving highly successful multiple agency collaborations in a cross-cultural environment: Experiences and lessons from Dhimurru Aboriginal Corporation and partners. *Ecological Management & Restoration, 13*, 42–50.
Ivison, D., Patton, P., & Sanders, W. (Eds.). (2000). *Political theory and the rights of indigenous peoples* (2nd edn.). Cambridge University Press.
Jackson, S. (2006). Compartmentalising Culture: The articulation and consideration of Indigenous values in water resource management. *Australian Geographer, 37*, 19–31.
Jackson, S., Porter, L., & Johnson, L. C. (2017). *Planning in Indigenous Australia: From imperial foundations to postcolonial futures*. Taylor & Francis Group.
Jarvis, D., Stoeckl, N., Addison, J., Larson, S., Hill, R., Pert, P., & Lui, F. W. (2018). Are Indigenous land and sea management programs a pathway to Indigenous economic independence? *Rangel Journal, 40*, 415. https://doi.org/10.1071/RJ18051
Kahane, D., Weinstock, D., Leydet, D., & Williams, M. (Eds.). (2010). *Deliberative democracy in practice* (New ed.). University of British Columbia Press.
Mabo v State of Queensland (No. 2) HCA 23, 1992.
Mackie, K., & Meacheam, D. (2016). Working on country: A case study of unusual environmental program success. *Australasian Journal of Environmental Management, 23*, 157–174. https://doi.org/10.1080/14486563.2015.1094752
Maddison, S., & Denniss, R. (2009). *An Introduction to Australian public policy*. Cambridge University Press.
Moorcroft, H., Ignjic, E., Cowell, S., Goonack, J., Mangolomara, S., Oobagooma, J., Karadada Dianna, R., Waina, N. (2012). Conservation planning in a cross-cultural context: The wunambal gaambera healthy country project in the Kimberley, Western Australia. *Ecological Management & Restoration, 13*(1), 16–25. https://doi.org/10.1111/j.1442-8903.2011.00629.x
Moran, M. (2016). *Serious Whitefella stuff: When solutions became the problem in indigenous affairs*. Melbourne University Press.
Moreton-Robinson, A. (2015). *The white possessive: Property, power, and indigenous sovereignty*. University of Minnesota Press.
Nakata, S., & Maddison, S. (2019). New collaborations in old institutional spaces: Setting a new research agenda to transform Indigenous-settler relations. *Australian Journal of Political Science, 54*, 407–422.
Neale, T. (2017). *Wild articulations*. University of Hawai'i Press.

Neale, T., Carter, R., Nelson, T., & Bourke, M. (2019). Walking together: A decolonising experiment in bushfire management on Dja Dja Wurrung Country. *Cultural Geographies, 26*, 341–359.
Owens, J., & Wilson, L. (2010). Noel Pearson brands Wild Rivers law "colonialism." *The Australian*.
Perry, M. (2014). Characterising native title rights: a desert rose by another name...
Pleshet, N. (2018). Caring for country: History and alchemy in the making and management of indigenous australian land. *Oceania, 88*, 183–201.
Putnis, A., O'Leary, P., Leach, A., Ings, E., & See, P., (2021). Strong on country: Sustaining success in Indigenous land and sea management in Australia. *Country Needs People*.
Reedy, J., Orr, R., Spicer, P., Blanchard, J. W., Hiratsuka, V. Y., Ketchum, T. S., Saunkeah, B., Wark, K., & Woodbury, R. B. (2020). Deliberative democracy and historical perspectives on American Indian/Alaska native political decision-making practices. *Humanities and Social Sciences Communications, 7*, 1–11.
Rose, D. B. (1996). *Nourishing terrains: Australian aboriginal views of landscape and wilderness*. Australian Heritage Commission.
Roughley, A., & Williams, S., (2007). *The engagement of Indigenous Australians in natural resource management: key findings and outcomes from land & water australia and the broader literature* (Final Report). Land & Water Australia.
Russell, P. H. (2006). *Recognising aboriginal title: The Mabo case and indigenous resistance to English-settler colonialism*. UNSW Press.
Samuel, G. (2020). *Independent review of the EPBC act* (Final Report). Department of Agriculture, Water and the Environment.
Sanders, W. Australian National University and Centre for Aboriginal Economic Policy Research. (2002). *Towards an indigenous order of Australian government: rethinking self-determination as indigenous affairs policy*. Centre for Aboriginal Policy Research.
Schirmer, J., Dare, M., & Ercan, S. A. (2016). Deliberative democracy and the Tasmanian forest peace process. *Australian Journal of Political Science, 51*, 288–307. https://doi.org/10.1080/10361146.2015.1123673
Schultz, R., & Cairney, S. (2017). Caring for country and the health of aboriginal and Torres Strait Islander Australians. *The Medical Journal of Australia, 207*, 8–10.
Smyth, D. (2006). Indigenous protected areas in Australia. *Parks: International Journal for Protected Area Managers, 16*.
Steiner, J. (2012). *The foundations of deliberative democracy: Empirical research and normative implications*. Cambridge University Press.
Stephenson, M. A., & Ratnapala, S. (Eds.). (1993). *Mabo a judicial revolution: The aboriginal land rights decision and its impact on Australian law*. University of Queensland Press.
Stoeckl, N., Grainger, D., Esparon, M., Farr, M., & Larson, S. (2016). *Economic values and Indigenous protected areas across Northern Australia*. James Cook University.
Storrs, M., & Cooke, P., (2001). Caring for country: The development of a formalised structure for land management on Aboriginal lands within the Northern Land Council region of the Northern Territory 7.
Strakosch, E. (2019). The technical is political: Settler colonialism and the Australian Indigenous policy system. *Australian Journal of Political Science, 54*, 114–130. https://doi.org/10.1080/10361146.2018.1555230
Strelein, L. (2009). *Compromised jurisprudence: Native title cases since Mabo* (2nd ed.). Aboriginal Studies Press.
Sullivan, P. (2011). *Belonging together: Dealing with the politics of disenchantment in Australian Indigenous affairs policy*. Aboriginal Studies Press.
Sullivan, P. (2013). Disenchantment, normalisation and public value: Taking the long view in Australian indigenous affairs. *The Asia Pacific Journal of Anthropology, 14*, 353–369. https://doi.org/10.1080/14442213.2013.804871
Taylor, B., Robinson, C., & Lane, M. B. (2010). *Contested country: Local and regional natural resources management in Australia*. CSIRO Publishing.

Uhr, J. (1998). *Deliberative democracy in Australia: The changing place of parliament*. Cambridge University Press.

Vincent, E., & Neale, T. (2017). Unstable relations: A critical appraisal of indigeneity and environmentalism in contemporary Australia. *The Australian Journal of Anthropology, 28*, 301–323.

Walker, R., Jojola, T., & Natcher, D. (2013). *Reclaiming indigenous planning*. McGill-Queen's University Press.

Warren, M. E. (2009). Governance-driven democratization. *Critical Policy Studies 3*(1), 3–13. https://doi.org/10.1080/19460170903158040

Webber, J., (1999). Native Title as Self-Government. *UNSW Law Journal, 22*.

Weir, J., Stacey, C., & Youngetob, K. (2011). *The benefits associated with caring for country—literature review*. Australian Institute of Aboriginal & Torres Strait Islander Studies (AIATSIS).

Western Australia v Ward [2002] HCA 28, (2002).

Western Australia v Ward and Ors (2000) 170 ALR 159, (2000).

Wik Peoples v The State of Queensland & Ors; The Thayorre People v The State of Queensland & Ors [1996] HCA 40, 1996.

Woodward, E., Hill, R., Harkness, P., & Archer, P. (2020). *Our knowledge, our way in caring for country: Indigenous-led approaches to strengthening and sharing our knowledge for land and sea management. Best Practice Guidelines from Australian experiences*. NAILSMA and CSIRO.

Wunambal Gaambera Healthy Country Plan (2010).

Wyatt, K., & Ley, S. (2020). Funding certainty for Indigenous Rangers, Joint Media Release (Document). https://ministers.pmc.gov.au/wyatt/2020/funding-certainty-indigenous-rangers

Young, E., Ross, H., Johnson, J., & Kesteven, J. (1991). *Caring for country: Aborigines and land management*. Australian National Parks and Wildlife Service.

Young, I. M. (2000). *Inclusion and democracy* (2nd ed.). Oxford University Press. https://doi.org/10.1093/0198297556.001.0001

# Chapter 5
# Future-Proofing Indigenous Self-Determination in Health: Goals, Tactics, and Achievements of Aboriginal Community Controlled Health Organisations in New South Wales, Australia

**David Coombs**

**Abstract** In this chapter I examine some of the ways that Aboriginal Community Controlled Health Organisations (ACCHOs) have responded to recent federal policy turmoil in Australia. I highlight ACCHOs as contemporary examples of Indigenous self-determination in health; these grassroots organisations are persisting in their work, designing community-owned responses to complex financial and political challenges. Drawing on data generated through a series of semi-structured interviews with CEOs and managers from ACCHOs in New South Wales, I explore these First Nations organisations' innovative and pragmatic responses to top-down policy changes. The interview data was generated with interviewees in urban, regional, and remote contexts. A thematic analysis of the data highlighted some of the on-the-ground effects of recent changes to Indigenous affairs funding and policy at the federal level. The policies in focus here, all of which were developed by the Liberal-National Coalition after its 2013 election win, are the Indigenous Advancement Strategy (IAS), the Indigenous Australians' Health Programme (IAHP) and the Primary Health Network (PHN). The federal government developed and implemented all three policies without substantial input from First Nations leaders or ACCHOs. The academic and grey literature demonstrates how these policy developments have threatened ACCHOs' access to primary health funding and imposed additional administrative and accountability requirements. In interviews, ACCHO

A note on terminology: Below, I use the terms 'Indigenous', 'First Nations', and 'Aboriginal and Torres Strait Islander' in reference to the First Peoples and Traditional Owners of the continent now known as Australia, including the islands of the Torres Strait. These three terms are used interchangeably. In my language choices I seek to be inclusive but also to be respectful of the diversity that exists within and between Indigenous peoples.

D. Coombs (✉)
School of Education, University of New South Wales, Kensington, NSW, Australia
e-mail: David.coombs@unsw.edu.au

The Australia and New Zealand School of Government (ANZSOG), Carlton, VIC, Australia



N. Moodie and S. Maddison (eds.), *Public Policy and Indigenous Futures*, Indigenous-Settler Relations in Australia and the World 4,
https://doi.org/10.1007/978-981-19-9319-0_5

leaders explained how their organisations have developed and implemented adaptive strategies to mitigate the impact on health services and outcomes. These strategies include diversifying income streams to reduce reliance on Commonwealth grant funding; partnering with health institutions at the state-government level to overcome Commonwealth funding austerity; and engaging in entrepreneurial activity. The chapter concludes that in the face of sustained fiscal pressure and regulatory constriction from Commonwealth policymakers, ACCHOs in NSW are resisting domination through innovation and pragmatism. However, in the process, some of these organisations are tending towards a more neoliberal mode of operation. This case study from the health sector illustrates how Indigenous peoples continue to exert their agency and slip through the cracks in the settler colonial edifice (Ford 2011; Simpson, A. 2014). It also shows how Indigenous organisations channel their policy and business acumen to survive in an increasingly competitive Indigenous health 'market'.

**Keywords** ACCHOs · Aboriginal health · Self-determination · Aboriginal community control · Indigenous health policy · New South Wales · Medicare

## 5.1 Introduction

Contemporary studies of Indigenous health in Australia, of which this chapter is one, are set against the backdrop of persistent 'gaps'[1] between First Nations and settler health outcomes and experiences. The higher rates of First Nations mortality and morbidity compared to non-Indigenous rates, constitute the fundamental policy and moral 'problem' of Indigenous health. In 2020, the Australian Government reported that the goal of closing the life expectancy gap between Indigenous and non-Indigenous Australians by 2031 was 'not on track' (Australian Government, 2020). Recent data shows a life expectancy gap of 8.6 years for men and 7.8 years for women (AIHW, 2020). This ongoing policy failure has led Indigenous organisations and advocates to call for a new engagement between governments and Indigenous peoples across Australia (see, for example, Coalition of Aboriginal and Torres Strait Islander Peak Organisations and all Australian Governments, 2020).

To improve Indigenous health, many First Nations people and organisations argue the Commonwealth government should increase its investment in and support for Indigenous self-determination and Aboriginal Community Control (ATSIPO, 2016, pp. 8–13; Lowitja Institute, 2021). In the health sector, Indigenous leaders and researchers have been advocating tirelessly for funding increases and policy support for Aboriginal Community Controlled Health Organisations (ACCHOs). In a recent report for the Close the Gap Steering Committee (2020, p. 17), the Lowitja Institute, charged that:

[1] The Closing the Gap targets and associated policy framework have been critiqued extensively since their introduction in 2008. See, for example, Altman (2009) and Pholi et al. (2009).

> Despite the innovation and resilience evident within their history, ACCHOs and other community controlled sectors have long struggled to make ends meet under inequitable and burdensome funding frameworks and processes.

This chapter is structured as follows: First, I examine the origins and unique features of ACCHOs, and outline the funding and policy arrangements in which they operate. I also explain the significance of the IAS, IAHP and PHN policy changes. Subsequently, I explain my research ethics, methods, and methodology. I then review the literature that deals with ACCHOs' contributions to Indigenous health and wellbeing. Later, I present my analysis of the qualitative interview data that forms the evidentiary basis of this chapter, highlighting four distinct themes in the interview responses. This is followed by a discussion of the significance of ACCHOs as contemporary examples of Indigenous self-determination, highlighting their capacity to enact community-owned responses to complex policy challenges. I conclude that ACCHOs are crucial sites of Indigenous health service development and delivery, and that their community embeddedness and health policy expertise should be acknowledged, respected, and appropriately resourced.

## 5.2 What Are ACCHOs?

ACCHOs deliver comprehensive and culturally appropriate healthcare to Aboriginal and Torres Strait Islander peoples. They emerged before Australia's systems of universal public health insurance were established (Duckett & Wilcox, 2015, p. 208). As Gary Foley explains, ACCHOs were a response to institutional racism so severe that 'in some cases...Aboriginal people would literally rather die than be subjected to degrading humiliating treatment at the hands of non-Aboriginal health workers' (Foley, 1991, p. 4). The first ACCHO, established in Redfern in 1971, was a 'self-help project, conceived, created and controlled by Indigenous people' (Foley, 2009, p. 16). Today, ACCHOs across Australia deliver a range of health-promotion, illness-prevention, and social programs. First Nations peoples own, run, and oversee their ACCHOs (Campbell et al., 2017; Grant et al., 2008, p. 8).

Scholars have argued that ACCHOs are a practical embodiment of Indigenous self-determination (Davis, 2013; Rowse, 2012, p. 101), giving Aboriginal people a say on what their health services do and how. As one of the Redfern Aboriginal Medical Service's (AMS) foundational members, Gary Foley maintains 'the basic philosophical principle that guides our work is "self-determination through community control"' (1991, p. 4). One common factor across the ACCHO sector is the Aboriginal community-controlled governance structure (Larkins et al., 2006, p. 2). For a health service to be Aboriginal community-controlled it must be initiated by and based in a local Aboriginal community and be governed by an Aboriginal board of directors elected by that community (NACCHO, 2011, p. 5). In addition, the services an ACCHO offers must be 'holistic and culturally appropriate' (NACCHO, 2011, p. 5). Aboriginal community-control, holism, and cultural appropriateness are at the heart of the ACCHOs' philosophy of healthcare.

## 5.3 Enduring Financial Insecurity

Funding challenges have long plagued the 'Indigenous sector' (Rowse, 2005), of which ACCHOs are a part. Almost two decades ago, Briskman (2003, p. 165) related Aboriginal organisations' concerns about 'the inadequate amounts of money and cutbacks [and] restrictive and bureaucratic accountability mechanisms imposed by governments'. More recently, Dwyer et al. (2009) detailed the 'crushing burden of regulation' that governments impose on ACCHOs. These researchers also highlighted the complex, short-term, and fragmented funding arrangements on which ACCHOs subsist. Subsequent research highlighted the ongoing funding and reporting challenges that ACCHOs continue to navigate (Martini et al., 2011; see also Dwyer et al., 2011). Despite some contract extensions, funding consolidations, and reporting efficiencies made by the federal Department of Health (Lowitja Institute, n.d., p. 2; Moran et al., 2016, p. 367), ACCHOs are still struggling to obtain sufficient and flexible funding from governments.

ACCHOs do not subsist on grant funding alone. They also generate income via the Medicare system. Since 1996 ACCHOs have been able to 'bulk bill' through Medicare (Bell et al., 2000, p. 82). ACCHOs' access to Medicare funding was increased in 2001 when the Howard government made changes to the Health Insurance Act 1973. This allowed ACCHOs to claim Medicare rebates in addition to receiving grant funding (Grant et al., 2008, p. 15). More recently, the addition of Indigenous-specific items to the Medicare Benefits Schedule, such as Indigenous health checks and chronic disease care plans, has seen ACCHOs' Medicare revenue increase (Moran et al., 2014, pp. 21–2). In 2014 it was reported that ACCHOs received 'about $700 million for Indigenous-specific services plus a Medicare component of about $45-50 million' (Russell, 2014, 17). Some ACCHOs are actively seeking to maximise their Medicare revenue to ensure their financial security, as the interview data below demonstrates. Several ACCHOs' publicly available financial statements show that Medicare revenue is becoming a larger proportion of their total revenue (see for example Awabakal, 2018, p. 6; Bullinah AHSAC, 2015, p. 14). This is an important development as Medicare revenue can be expended in a more flexible manner than grant revenue, as it is not tied to particular outcomes or activities once generated through service provision. Nevertheless, Medicare funding remains a small proportion of ACCHOs' total revenue stream and ACCHOs are still heavily reliant on Commonwealth grants for Indigenous service provision.

In the 2014 budget the conservative Liberal-National Coalition under Prime Minister Tony Abbott, introduced three major Indigenous Affairs funding reforms: the Indigenous Advancement Strategy (IAS), the Indigenous Australians' Health Programme (IAHP), and the replacement of Medicare Locals with Primary Health Networks (PHNs). These three policies promote competition amongst service providers, based on the assumption that 'contestability' will drive innovation and efficiency (Productivity Commission, 2017, p. 865). The Abbott government's first budget also pursued fiscal austerity, which is closely associated with neoliberalism (Schrecker & Bambra, 2015). Handing down the budget, Federal Treasurer Joe

Hockey, photographed that night with a fat cigar in hand, disclosed that more than half a billion dollars had been cut from federal Indigenous Affairs, including $165.8 million over four years from Indigenous health programs (Australian Government, 2014; Russell, 2014). It was reported that $112,884 of this had come from cuts to ACCHOs (NACCHO, 2015). This was a significant reduction especially in light of the sector's persistent funding shortfalls (Alford, 2014; Couzos & Delaney Thiele, 2007, p. 523; Grant et al., 2008, p. 19).

The IAS, IAHP and PHNs are broadly similar in their intent and all three policies have the potential to undermine the financial security and legitimacy of the ACCHO sector. All three policies introduce a level of competitive tendering for government grants and contracts, privilege corporate models of governance and accountability, and prioritise 'value for money', efficiency and effectiveness (DoH, 2016; PM&C, 2014). These are the concepts and techniques of New Public Management (Hood, 1991, pp. 4–5). The IAS, IAHP and PHNs signalled that the federal government remains unsupportive of Indigenous self-determination, preferring what could be characterised as 'market determination'. I argue that these three policies introduced a new level of 'marketisation' to the Indigenous health sector (Meagher & Goodwin, 2015, p. 2).

Indigenous peoples and their representative and service-delivery organisations have heavily criticised recent Indigenous Affairs policy developments such as the IAS (see, for example, ATSIPO, 2016; ATSISJC, 2016; NACCHO, 2015; Senate FPARC, 2016). Indigenous leaders and health policy scholars are also still urging governments to afford ACCHOs greater autonomy and resources. Nevertheless, funding is still piecemeal and short-term. In these circumstances, many ACCHOs have developed strategies that guarantee themselves a level of independence from governments, in the hope that this might guarantee their survival and expansion into the future. This chapter is an exploration of these survival strategies.

Given this context, this chapter examines how ACCHOs are future-proofing themselves in response to ongoing policy and funding challenges. The chapter contributes new knowledge about the on-the-ground effects of federal Indigenous health policy decisions, and details the community-controlled responses to these nascent challenges. There is very little academic literature that focuses on how the IAS, IAHP, or PHN have affected ACCHOs (notable exceptions include Coombs, 2018; Strakosch, 2019). Even less has been written about how ACCHOs are responding to these emerging policy and funding threats. By examining how ACCHOs are faring amidst these new policy arrangements, I argue that the effectiveness of government policies can be better understood. I also argue that by examining examples of ACCHOs' innovation the case for increased Indigenous self-determination in health becomes stronger.

By engaging with evidence generated through interviews with senior ACCHO staff, I highlight ACCHOs' resilience and the innovations that they have developed in the face of funding instability. The data shows that ACCHOs feel under increasing pressure to corporatise and become more business-like in their operations. Interviewees explained the ways in which they are attempting to strengthen their community-controlled governance model, while also guaranteeing their organisations' stability.

This chapter focuses on the adaptive strategies deployed by ACCHOs so that other Aboriginal community-controlled organisations can learn from their experiences. Indigenous organisations in other policy areas may be able to learn from ACCHOs' stories in health. Policy makers may also be interested in this analysis, gaining a deeper appreciation of ACCHOs' strengths and innovative capacities. Additionally, this chapter may serve as evidence to justify greater Indigenous self-determination in health.

The starting premise of this chapter is that ACCHOs carry out a unique and integral function in the health system. I argue that the community-controlled governance model is what makes ACCHOs effective, and that if ACCHOs were to disappear then Indigenous health and wellbeing would be severely affected. Self-determination is also a human right under The United Nations Declaration on the Rights of Indigenous Peoples (UNDRIP), to which Australia is a signatory. ACCHOs have demonstrated their clinical effectiveness and their ability to tackle the complex socioeconomic causes of Indigenous health inequity. The chapter now turns its attention to the evidence of ACCHOs' effectiveness.

## 5.4 Ethics, Methodology, Method

Two primary ethical considerations guided this project: 'privileging Indigenous voices' (Rigney, 1999, p. 166); and taking a 'strengths-based approach' to research with and about Indigenous peoples (Fogarty et al., 2018). I adopted a decolonising methodology following the principles outlined by Linda Tuhiwai-Smith (1999). The data gathering and analytic methods I used were semi-structured interviews and thematic analysis, drawing on the work of Pope, Ziebland and Mays (2007), and Ezzy (2003). The data and analysis for this chapter is drawn from a larger project where I conducted 22 interviews with 29 individual participants at 18 ACCHOs across NSW. Interviews were conducted between July 2017 and March 2018, and took place in urban, rural, and remote parts of NSW. All participants have been deidentified and given pseudonyms. This research received ethics approval from the Ethics Committee of the Aboriginal Health and Medical Research Council of New South Wales (project number 1225/16). For this chapter I also conducted desk-based research, which consisted of a literature review of the evidence on ACCHOs' successes and contributions to the Australian health system.

## 5.5 ACCHOs' Achievements in Indigenous Health

There is already considerable evidence demonstrating ACCHOs' many achievements in Indigenous health. The growing body of evidence that shows how ACCHOs generate health and social benefits sits in stark contrast with the Australian government's ongoing resistance to allowing ACCHOs more stability and autonomy.

Firstly, research has highlighted how ACCHOs' interventions target the underlying structural causes of Indigenous health inequity. Pearson et al. (2020) found that ACCHOs across Australia deliver a broad range of health and social services that directly address the social determinants of health (see also Campbell et al., 2017, F). These are the social, political and economic 'conditions in which people are born, grow, live, work and age' (CSDH, 2008, p. 1). There is a long history of international research showing how the social determinants of health, rather than individual or biological causes, are the primary drivers of human health and illness (see for example AIHW, 2018; CSDH, 2008; Marmot, 2005, 2011; Wilkinson & Marmot, 2006). The Australian Institute of Health and Welfare (2018, p. 6) has calculated that five social determinants related to employment, qualifications, schooling, housing, and income explain 53 per cent of the health gap between Indigenous and non-Indigenous health outcomes. ACCHOs improve First Nations peoples' socioeconomic conditions in these areas. For example, ACCHOs act as 'local employment hubs' for First Nations peoples (Lovett, 2014, p. 45). The ACCHO sector is the largest single employer of Aboriginal and Torres Strait Islander peoples in Australia (Alford, 2014, p. 11), employing over 3,300 First Nations workers in 2017–18 (NACCHO, 2018, p. 43). 56 per cent of the ACCHO workforce is Indigenous (NACCHO, 2018, p. 43). By creating Indigenous employment opportunities and boosting Indigenous household income, ACCHOs are making a direct impact on two important determinants of Indigenous health.

Secondly, the few empirical studies that have compared ACCHOs' services with mainstream general practice and other primary health care services suggest that that ACCHOs offer highly effective and holistic health services. For example, Freeman et al. (2016, p. 99) found that the Central Australian Aboriginal Congress offered a broader range of health services (at no additional cost to the clients) compared to state-managed and nongovernmental services. Hurley et al. (2010) identified ACCHOs as the organisations in Australia that most faithfully implement the internationally recognised model of Comprehensive Primary Health Care. Similarly, Larkins et al. (2006) found that ACCHOs' clinicians manage a greater number of health problems per consultation than mainstream general practitioners who treat Indigenous patients. In an earlier study, Thomas et al. (1998) made similar findings, concluding that ACCHOs' consultations were more complex and managed a greater number of problems than mainstream general practice. Two studies comparing the clinical outcomes achieved by ACCHOs with those of mainstream general practices concluded that ACCHOs' outcomes were as good or better than mainstream services (Jan et al., 2004; Panaretto et al., 2005). ACCHOs have also been shown to deliver valuable health education (Baba et al., 2014, p. 6) and to support clients in their follow-up treatment (Panaretto et al., 2014). The available evidence leads to the conclusion that ACCHOs' services deliver broad health benefits to their communities, and that they often outperform their mainstream counterparts in this endeavour. Nevertheless, there is a need for further study and evaluation of ACCHOs' services and their health impacts.

Thirdly, as well as the studies that have highlighted ACCHOs' achievements in health, other scholars have advanced persuasive arguments heralding ACCHOs' cultural and political strengths. Regarding cultural strengths, scholars and healthcare practitioners have argued that ACCHOs prioritise 'culturally appropriate' or 'culturally safe' care in order to combat and undo some of the damage done by institutional racism in mainstream health settings (Best & Fredericks, 2014; Coffin & Green, 2017; Taylor & Guerin, 2010). Culturally safe healthcare is concerned with reducing power inequalities, stereotyping, and biases, and increasing critical consciousness in both individual healthcare practitioners and healthcare organisations. Curtis et al. (2019) maintain that cultural safety:

> requires individual healthcare professionals and healthcare organisations to acknowledge and address their own biases, attitudes, assumptions, stereotypes, prejudices, structures, and characteristics that may affect the quality of care provided. (p. 14)

Institutional racism is now well-recognised as a major problem that diminishes Indigenous peoples' experiences of the Australian healthcare system (Bourke et al., 2018; Durey, 2010; Henderson et al., 2007, p. 8; Paradies et al., 2008, p. 9). It can be understood as the unconscious biases and norms that inform how health practitioners interact with and care for patients (van Ryan & Fu, 2011). Institutional racism can deter Indigenous patients from accessing healthcare (Awofeso, 2011; Baba et al., 2014; DoH, 2017a, p. 23; Osborne et al., 2013; ) and make it less likely that Indigenous patients will receive appropriate medical care (Cunningham, 2002; Hall et al., 2007; Moore et al., 2014). Experiencing racism in healthcare settings has also been associated with heightened levels of psychological distress amongst Indigenous patients (Kelaher et al., 2014). ACCHOs' ongoing campaign to eliminate institutional racism from healthcare and their leadership in the delivery of culturally safe care to Indigenous peoples (Baba et al., 2014; Campbell et al., 2017; Durey, 2010), makes them a unique and integral part of the Australian health system.

The Australian government is promoting competition and contestability in Indigenous health funding processes; they are the key elements of this recent application of marketisation to Indigenous Affairs (Meagher & Goodwin, 2015, p. 5; see also Thornton et al., 2012). According to Australia's Productivity Commission: '[c]ontestability means that a provider of human services faces a credible threat of replacement if it underperforms' (2017, p. 65). Contestability poses a serious threat to ACCHOs' survival. It means that Indigenous organisations must compete with non-Indigenous service providers in order to win Indigenous health funding. Under this funding model, the federal government retains control over decision-making, and ultimately decides what services are offered, to whom and how. Nevertheless, ACCHOs and other Indigenous organisations have responded pragmatically and innovatively to these challenges. Below, I present evidence from interview data that illustrates some of the strategies ACCHOs are adopting to ensure the ongoing delivery of critical health and social services to their communities, in spite of the federal funding insecurity.

## 5.6 Data Analysis and Findings

### *5.6.1 Becoming Medicare Machines*

Many ACCHO representatives outlined how, in this difficult funding environment, their organisations were trying to pivot away from federal grant funding and make greater use of Medicare revenue. Some interviewees also explained that this strategy was a way of lessening their organisations' dependence on government grant funding. Evidence from interviews shows that several ACCHOs in NSW are aligning their services with the financial incentives of the Medicare Benefits Schedule. In many cases, interviewees stated that the Medicare funding system affords their ACCHOs some flexibility in service delivery, as they are able to divert any surplus Medicare revenue towards the under-funded aspects of their service. For example:

> … when we lose a program, we have to figure out how we keep that program going, and generally that's through our Medicare revenue. That's what we use that [revenue] for, to pay for programs that we don't get funded for. [But we provide it anyway] because it's a community need. (Lizzie)

Similarly, another ACCHO uses Medicare to supplement the funding of a youth program:

> … we just signed off on two Youth Development Officers – from our Medicare funds – for a pilot of 12 months, because you can't go past what you're going to generate in Medicare in the next year. So we'll fund those positions from our Medicare [revenue] because that's what the community wants … I think we get 70 grand from Family and Community Services for a youth position. It's not even enough to put anyone one who's got any skills to do anything, you know. … It's crap. When you take your on-costs out of it there's nothing left. So we'll fund that and … they'll have a great impact on access to our services for youth … (Rosemary)

Sam's statement, below, summarises the costs and benefits of making greater use of Medicare funding:

> Well there are AMSs that feel that they've been pushed by government into becoming a Medicare machine to generate income and some of those AMSs have allowed other people in the general community to come into their Medical Services so they can generate Medicare. They've been squeezed and pushed into that environment because ultimately in an AMS, doctors generally tell me that they've got to have long consultations with individuals because they don't just come there with one illness they come there with multiple illnesses. They say if they give a short consultation and they're only treating one illness, that doesn't do the patient justice. (Sam)

One ACCHO has even employed a Medicare officer to ensure that GPs are claiming the total amount of Medicare revenue they are entitled to. According to Charlie, the CEO of this ACCHO, the Medicare officer 'pays for herself' via the revenue she generates through auditing Medicare claims and receipts. The above interview responses, and others like them from the rest of the interview cohort, show how Medicare revenue is becoming an increasingly valuable income stream for ACCHOs who are seeing their grant funding diminish or be destabilised. However, this does not come without risks, and some interviewees feel as though patient care may be compromised because of the increasing pressure to make money through the fee-for-service MBS.

### 5.6.2 *Mitigating Financial Risk*

Other ACCHOs are taking a different approach to manage their financial risk. A minority of ACCHO CEOs stated that their services had adopted a strategy of income stream diversification in response to the possibility that they might lose some of their government funding. For example, Harry states:

> ... when I first started here the budget was 750,000 dollars and that was it. And that was to run the service. That was Commonwealth money. And ... over time we've increased it to diversify the income streams that are coming in, which means ... if one of those funding streams fell apart and we weren't re-funded, there are other opportunities... It's a risk environment but because we've diversified, we've not put all our eggs in one basket. We're looking at how to ensure the service continues even though we might lose a funding source from time to time. (Harry)

He continues,

> Because there's such a period of change that's occurring around structure and funding and guidelines and Medicare, I think what we need to do is work out ... how we can future proof the organisation. Because if things get tougher and you don't do that then I think there's a difficulty. I don't think you can rely on government bodies to make some of those decisions, you've got to do that for yourself. (Harry)

This interviewee and the other ACCHO representatives who made similar comments highlight the agency that some ACCHOs exercise, even in the face of an uncertain future of government funding for Aboriginal health.

### 5.6.3 *Two-Way Learning: Sharing Knowledge and Resources*

Around one third of the interviewees emphasised the importance of sharing complementary skills and knowledge between mainstream health actors, such as hospitals and Local Health Districts, and ACCHOs. The CEO of an inner-regional ACCHO, who also has experience working in Central Desert communities, gave the most detailed explanation of this exchange, drawing on the Aboriginal concept of the *Malparara* Way or 'the strength of working together' (see also NPY Women's Council, 1999). During a discussion about the need for closer partnerships between ACCHOs and mainstream organisations, he posited:

> I think [this ACCHO] would be naïve to think that we can survive without those effective partnerships. And they need to be two-way learning. That word is the *Malparara* way. It means you bring something to the table, I bring something to the table. Together we're

> better, and we will deliver better services, more functionally, more appropriately, more respectfully. … the *Malparara* way is a way of communicating effectively that's respectful to both skills that come to the table. … The mob get it because the Elders functioned on respectful communication, and if that falls over, the partnerships would fall over. (Tim)

Respectful communication and sharing resources are the essence of two-way learning and the key to successful partnerships, according to this non-Indigenous CEO. The clinical services manager from this same ACCHO described two-way learning using similar terms, but emphasised Aboriginal organisations' leading role:

> The role of [ACCHOs] is to lead, should be, to lead mainstream as to what is culturally appropriate for the people and what is not culturally appropriate for the people. And … it's a two-way learning: we can learn from them and they can learn from us. (Dane)

Other interviewees, while not specifically referring to the two-way learning concept, emphasised the way that people trained in the mainstream health system and ACCHOs could share knowledge for the betterment of Aboriginal health. For example, this CEO reflects on the learning of a doctor who has been working at her ACCHO:

> What we want is for him to go out into the bigger, brighter world and treat people, Aboriginal clients and other Aboriginal community members with the respect that we have taught him, you know. And his medicine for Aboriginal people should be really great. (Linda)

Another CEO described a similar dynamic in her ACCHO:

> And it's like, we contract a private psychologist, but we had him working out of our building for a couple of years beforehand, so he became culturally appropriate. So we're also educating service providers. (Amy)

While these examples relate to one side of the two-way learning process, the fact that ACCHOs contract and employ health professionals trained in mainstream institutions is evidence that they value the clinical knowledge they bring with them. A minority of other interviewees spoke of more formalised instances of two-way learning, via staff exchange programs between ACCHOs and hospitals. Facilitated by a partnership between an LHD and an ACCHO, Aboriginal and non-Aboriginal staff alike gain a better understanding of Aboriginal community-controlled and mainstream health contexts. The Aboriginal CEO of the ACCHO involved explains how he and his LHD partners have arranged for:

> non-Aboriginal staff to come and work in the AMS, and Aboriginal staff to work in the hospitals. So we become familiar with those environments, to break down their barriers and yeah, which creates more effective communication. (Sam)

Sam's testimony affirms that not only do staff exchanges and partnerships increase the level of clinical health services available to Indigenous clients, but they also make the mainstream health system more culturally safe. ACCHO representatives from a discrete Aboriginal community in an outer-regional area spoke hopefully about a similar staff exchange in their area, referring to a recent memorandum of understanding (MOU) between the local hospital and the ACCHO:

Natalie: And another thing too, [the MOU] will make the hospital more culturally appropriate. They'll come out here and do some work experience ... doctors, nurses, admin. We're going to sort of swap them... Our workers can get the mainstream experience, theirs will get the Indigenous health ... experience.

Me: So they'll learn a lot from you, for sure.

Amanda: And vice versa.

These interviewees point to the mutual benefits and knowledge gains that accrue to both parties to the agreement. Through partnership, ACCHOs gain some financial stability and can access a broader range of resources and personnel. Additionally, through two-way learning, ACCHOs may be able to effect cultural changes in the mainstream health system.

Other interviewees alluded to productive interactions between mainstream and Aboriginal community *knowledges* when discussing the ways that doctors and Aboriginal health workers could cooperate in caring for Aboriginal clients' health. While some interviewees contended that the skill set of Aboriginal Health Workers was not sufficiently valued in mainstream settings, a few interviewees argued that their skill sets were complementary, and that the relationship needed to be based on 'equal partnership'. One interviewee contended that:

> It goes back to respect, which goes both ways. Health workers and doctors should show mutual respect. Both have to make sure they are working together to heal the client. Doctors have formal training, but Aboriginal Health Workers have training in life experience. (Caitlin)

This senior Aboriginal manager also contended that Aboriginal Health Workers could facilitate effective communication between non-Aboriginal doctors and older Aboriginal clients, and help them to navigate cultural protocols surrounding 'men's and women's business'. While this interviewee emphasised that mainstream health professionals can learn from ACCHOs, other interviewees underlined the areas where ACCHOs could both learn from and make use of mainstream health professionals and their skills, particularly in relation to corporate governance.

### *5.6.4 Re-structuring ACCHOs Around Business Principles*

A final organisational survival strategy mentioned by some interviewees involved trying to run their ACCHOs more like a business in recent years. For a minority of interviewees, this decision was taken in order to make their organisation more financially independent:

> [W]e're really just looking at the commercial and business opportunities that the organisation needs to make to be viable because we don't want to be dependent on government funding. (Rosemary)

Similarly:

> The funding – you're always fearful ... that the government comes in and changes and cuts your funds... So that's how we've turned our thing around and [we] look to run like a business

> so that if it ever happens ... you've got that backup and you're viable without relying on their funding. (Dominique)

In a related way, other interviewees framed their decision to make their ACCHO more business-like as a response to Government pressure or incentives. Approximately one third of the interview cohort said that their services were under pressure to become more business-savvy. Most commonly, interviewees cited the funding reforms associated with the Indigenous Advancement Strategy (IAS) and the PHNs, and the upcoming changes to the IAHP, as the source of this pressure. Harry articulated this pressure as a government expectation':

> [A]ll funding bodies in my view are moving towards a business model of service not a community-based model of service and so the expectations are that you will run as a business. So you have to have things in place that will allow you to function in that environment and the structure of the board needs to understand that, you know, it's a business in its purest sense. And yes we're providing community service but we have to run like a business. (Harry)

Additionally:

> Government's been saying this for the last six years: that we *are* a business. (Charlie)

And again:

> [W]e are being pushed to be more business-savvy ... It's coming from the Department of Health, and from State Health actually. They're all looking at what we're making through Medicare, making sure that we're maximising Medicare. And that's impacting on, well from a clinician's point of view they feel it's impacting on patient care. (Rosemary)

This same interviewee, an urban-based ACCHO CEO, interpreted governments' preference for business-like ACCHOs as an economic decision. However, she also argued that governments were overlooking valuable Aboriginal community obligations:

> I think the government's main concern is about how their money's being spent, and what's the outcome from their investment. And I get all that but there is also the accountability and responsibility back to community for the services you deliver. If the government were that interested in what was happening [they'd] go and talk to the community. They don't do that. They're not interested in that. (Rosemary)

A number of ACCHO representatives explained that their services had developed a commercial side to their operations, in order to generate more revenue that could then be used to expand services to community. As Tristan, the CEO of a remote-located ACCHO, outlines:

> So we're a pretty big organisation and we're a lot different to most AMSs... we have other businesses that we own and run and we use ... that income to support the AMS in delivering its services.... so we've got four clinics: two AMS clinics and we've got two private practices. We own a gym, we own a funeral home. Yeah so we've got a bit of everything. We auspice the management of ... [another] Aboriginal Medical Service, we run it as well. All up I think we've got about 75 staff on the books. So we're the largest private, non-government organisation employer in [the] shire so yeah we employ about 15 doctors right across the organisation, so yeah we're pretty big. (Tristan)

Similarly, other interviewees explained that their ACCHOs offer GP services to non-Aboriginal patients, generating Medicare revenue that can be used to subsidise Aboriginal-specific services and programs, which often do not generate revenue or attract adequate funding. In order to generate sufficient income to maintain a broad range of services and programs some ACCHOs have to provide many instances of care to the non-Aboriginal population. In effect:

> We have about 11,000 people on our books. Just over 3000 are Aboriginal people. So we have Aboriginal-specific programs, so obviously our GP clinics and our nurses, so the medical clinic is open to everyone. We're probably the only bulk-billing service in town, that fully bulk bills. And then we have other services that are Aboriginal-specific so they're only open to Aboriginal people and their non-Aboriginal spouses. Like the drug and alcohol program and a whole range of other programs that we've got here. (Gwen)

Gwen's ACCHO makes its services available to the Aboriginal and non-Aboriginal residents of the town. However, given that Gwen stated that her organisation's main goal was to 'to work towards Aboriginal people enjoying the same standard of health as non-Aboriginal Australians', the fact that her ACCHO services so many non-Aboriginal people appears to be a means of generating revenue rather than as a fundamental goal of the organisation. This indicates that the current funding model does not adequately compensate this ACCHO for its Aboriginal-specific programs and services, leading them to search for revenue through other, more 'mainstream' and commercial, areas of service provision. This pattern of service commercialisation has been observed across the Indigenous sector in recent decades (Hunt, 2008, p. 43), indicating systemic problems within both Commonwealth and state-level Aboriginal-specific funding models. Dane described a similar pattern of servicing increasing numbers of non-Aboriginal clients:

> … we see non-Indigenous people as well. … we're seeing that increase and we're seeing clients come from other areas in the region because of changes to Medicare, changes to the way that local doctors work. … We're structured around being a bulk-billing practice so we're seeing more and more of these clients come in. So at the moment, whilst our funding hasn't increased, we're taking on a whole 'nother (sic) portion of community … and it becomes a balancing act in that you are trying to provide health care for everybody [but] with a focus on Aboriginal and Torres Strait Islander people. (Dane)

This interviewee problematised the way that health funding structures can distract ACCHOs from their core business of providing healthcare to Aboriginal and Torres Strait Islander people.

## 5.7 Discussion: Defending Self-determination in Health

The above analysis has highlighted some of the survival strategies that ACCHOs have adopted in NSW. By prioritising the delivery of services listed on the Medicare Benefits Schedule, diversifying income streams, promoting two-way learning and partnerships, and becoming more business-like in their operations, ACCHOs in NSW have been able to future-proof their finances. These findings are similar to the findings of international studies that have looked at the benefits and practice of Indigenous self-determination in health (e.g. Carroll et al., 2021). These adaptations can be understood as pragmatic business decisions in response to challenging economic conditions. Alternatively, they can be understood as a stalwart defence of Aboriginal community control in a critically important policy domain. To conclude, I offer some reflections on the reasons why self-determination is so important to ACCHOs.

### *5.7.1 The Teleological Argument: Self-determination as a Way of Improving Indigenous Health*

Self-determination, or empowerment, can be understood as a means to an end, as a way of increasing Indigenous health outcomes. Empowerment's utility as a health promotion strategy has been theorised by public health researchers (e.g. Wallerstein, 2002; Labonte, 1994). Community empowerment has been defined as 'a social action process by which individuals, communities, and organizations gain mastery over their lives in the context of changing their social and political environment to improve equity and quality of life' (Wallerstein, 2002, p. 73). Other academic disciplines, such as community health psychology, also champion community empowerment for better health (Murray et al., 2004). This latter field of investigation aims to build community strength and challenge social oppression so that there is more 'personal and political awareness of the social constitution of health and illness' (Murray et al., 2004, p. 331). Murphy's (2014, p. 320) characterisation of self-determination as a collective capability' that enhances Indigenous freedom and wellbeing is another theoretical approach that links empowerment to health gains. Lardner suggests that 'self-determination may be a determining factor in enabling/disabling communities in crisis and in understanding resiliency' (2009, p. 93). She also contends that 'legitimacy', defined as responsible and accountable government, is the crucial variable that shapes how self-determination will affect community wellbeing. While the evidence base connecting self-determination to health is still largely theoretical, many interviewees in this study spoke of Aboriginal community control as an invaluable institution that makes ACCHOs accountable to their communities, affording local Aboriginal communities some control over their own health.

## 5.8 Conclusion: Self-determination as an End in Itself

There is a risk that such a utilitarian understanding of self-determination may reduce the significance of Indigenous people's aspirations for greater control over Indigenous lives, lands, and cultures. Indeed, Indigenous self-determination has been codified as an inalienable human right under international law. The UNDRIP asserts that Indigenous peoples 'have the right to autonomy or self-government in matters relating to their internal and local affairs, as well as ways and means for financing their autonomous functions' (article 4, p. 8). ACCHOs' recent adaptations can be understood as examples of Indigenous peoples exercising their inherent rights to self-determination, an important end in and of itself. These community-owned organisations have designed and implemented Indigenous policy responses to government policy upheavals. Based on this case study of Indigenous-run organisations fighting to preserve their autonomy in a key policy domain, this chapter concludes that the principle of Aboriginal community control in health is crucially important. ACCHOs have demonstrated significant adaptive capacity, resilience, and localised policy expertise. Future research is needed to better understand how ACCHOs' unique attributes and skills can be nurtured and enhanced towards the goals of Indigenous self-determination, health, and wellbeing.

## References

Aboriginal and Torres Strait Islander Peak Organisations [ATSIPO]. (2016). The redfern statement. In *National congress of Australia's first peoples*. http://nationalcongress.com.au/about-us/redfern-statement/

Aboriginal and Torres Strait Islander Social Justice Commissioner [ATSISJC]. (2016). *Social justice and native title report 2016*. Australian Human Rights Commission. https://www.humanrights.gov.au/our-work/aboriginal-and-torres-strait-islander-social-justice/publications/social-justice-and-2, accessed 14/01/2020.

Alford, K. (2014). Economic value of aboriginal community controlled health services. National Aboriginal Community Controlled Health Organisation. https://apo.org.au/node/39333

Altman, J. (2009). *Beyond closing the gap: Valuing diversity in indigenous Australia* (CAEPR Working Paper no. 54). Centre for Aboriginal Economic Policy Research. https://caepr.cass.anu.edu.au/research/publications/beyond-closing-gap-valuing-diversity-indigenous-australia

Australian Government. (2014). *Budget Paper No. 2, Budget (2014–2015), expenses measures: Indigenous affairs programmes—rationalisation*. Commonwealth of Australia. http://budget.gov.au/2014-15/content/bp2/html/bp2_expense-20.htm

Australian Government. (2020). *Closing the gap report 2020*. Commonwealth of Australia. https://ctgreport.niaa.gov.au/sites/default/files/pdf/closing-the-gap-report-2020.pdf

Australian Institute of Health and Welfare [AIHW]. (2018). 6.7: Size and sources of the Indigenous health gap. In *Australia's health 2018*. https://www.aihw.gov.au/getmedia/f1d1c24c-2966-4e06-9609-d7a099f6d3af/aihw-aus-221-chapter-6-7.pdf.aspx

Australian Institute of Health and Welfare [AIHW]. (2020). *Australia's health 2020*. Australian Government. https://www.aihw.gov.au/reports-data/australias-health

Awabakal. (2018). Annual Report 2017–2018, Awabakal Ltd, Wickham, NSW, available at: http://www.awabakal.org/Awabakal-Annual-Report2018-Web.pdf, accessed on 01 June 2020.

Awofeso, N. (2011). Racism: A major impediment to optimal Indigenous health and health care in Australia. *Australian Indigenous Health Bulletin, 11*(3), 1–13.

Baba, J. T., Brolan, C. E., & Hill, P. S. (2014). Aboriginal medical services cure more than illness: a qualitative study of how Indigenous services address the impacts of discrimination in Brisbane. *International Journal for Equity in Health, 13*(56). https://doi.org/10.1186/1475-9276-13-56

Bell, K., Couzos, S., Daniels, J., Hunter, P., Mayers, N., & Murray, R. (2000). Aboriginal community controlled health services. In Commonwealth Department of Health and Aged Care (Ed.), *General practice in Australia: 2000* (pp. 74–104). Commonwealth Department of Health and Aged Care, Canberra. http://www.kooriweb.org/foley/resources/pdfs/120.pdf

Best, O., & Fredericks, B. (Eds.). (2014). *Yatdjuligin: Aboriginal and Torres Strait Islander nursing and midwifery care*. Cambridge University Press.

Bourke, C. J., Marrie, H. M., & Marrie, A. (2018). Transforming institutional racism at an Australian hospital. *Australian Health Review*. https://doi.org/10.1071/AH18062

Briskman, L. (2003). 10: It's all about funding. In *The black grapevine: Aboriginal activism and the stolen generations*. The Federation Press, Annandale and Leichhardt

Bullinah Aboriginal Health Service Aboriginal Corporation [AHSAC]. (2015). Annual Report 2014, Ballina NSW, available at: http://bullinahahs.org.au/wp-content/uploads/2019/02/Annual-Report-2014-2015.pdf, accessed on 01 June 2020.

Campbell, M. A., Hunt, J., Scrimgeour, D. J., Davey, M., & Jones, V. (2017). Contribution of aboriginal community-controlled health services to improving aboriginal health: An evidence review. *Australian Health Review, 42*(2), 218–226. https://doi.org/10.1071/AH16149

Carroll, S. R., Cornell, S., & Jorgensen, M. (2021). Can a self-determination strategy improve Indigenous health care? Evidence from the United States'. In D. Smith, A. Wighton, S. Cornell, & A. V. Delaney (Eds.), *Developing governance and governing development: International case studies of Indigenous futures* (pp. 201–222). Rowman Littlefield International.

Coalition of Aboriginal and Torres Strait Islander Peak Organisations and all Australian Governments. (2020). National Agreement on Closing the Gaps. https://coalitionofpeaks.org.au/wp-content/uploads/2021/04/ctg-national-agreement-apr-21-1-1.pdf

Coffin, J., & Green, C. (2017). Decolonising Australian community development tools. In C. Kickett-Tucker, D. Bessarab, J. Coffin, & M. Wright (Eds.), *Mia Mia Aboriginal Community Development: Fostering cultural security* (pp. 73–90). Cambridge University Press.

Commission on the Social Determinants of Health [CSDH]. (2008). *Closing the gap in a generation: Health equity through action on the social determinants of health* (Final Report of the Commission on Social Determinants of Health). World Health Organization. https://www.who.int/social_determinants/thecommission/finalreport/en/

Coombs, D. (2018). Primary health networks' impact on aboriginal community controlled health services. *Australian Journal of Public Administration, 77*, S37-46. https://doi.org/10.1111/1467-8500.12357

Couzos, S., & Delaney Thiele, D. (2007). The international covenant on economic, social and cultural rights and the right to health: Is Australia meeting its obligations to Aboriginal peoples? *Medical Journal of Australia, 186*(10), 522–524.

Cunningham, J. (2002). Diagnostic and therapeutic procedures among Australian hospital patients identified as Indigenous. *Medical Journal of Australia, 176*(2), 58–62. https://doi.org/10.5694/j.1326-5377.2002.tb04284.x

Curtis, E., Jones, R., Tipene-Leach, D. *et al.* (2019). Why cultural safety rather than cultural competency is required to achieve health equity: a literature review and recommended definition'. *International Journal for Equity Health, 18*(174). https://doi.org/10.1186/s12939-019-1082-3

Davis, M. (2013). Community control and the work of the National Aboriginal Community Controlled Health Organisation: Putting meat on the bones of the UNDRIP. *Indigenous Law Bulletin, 8*(7), 11–14.

Department of Health. (2016). *Primary health networks: Grant programme guidelines*. Australian Government. http://www.health.gov.au/internet/main/publishing.nsf/Content/PHN-Program_Guidelines

Department of Health. (2017a). *My life my lead—Opportunities for strengthening approaches to the social determinants and cultural determinants of Indigenous health: Report on the national consultations*. Commonwealth of Australia.

Department of Health. (2017b). *Development of a funding model for the indigenous Australians' health programme*. Presented at AHandMRC Meeting Ground 14–16 June 2017b. http://www.ahmrc.org.au/media/resources/ahmrc-meetings/meeting-ground-2017b.html

Department of Prime Minister and Cabinet [PMandC]. (2014). *Indigenous advancement strategy guidelines*. Commonwealth of Australia. https://www.dpmc.gov.au/sites/default/files/publications/ias_guidelines.pdf

Duckett, S., & Wilcox, S. (2015). *The Australian health care system* (5th ed.). Oxford University Press.

Durey, A. (2010). Reducing racism in Aboriginal health care in Australia: Where does cultural education fit? *Australian and New Zealand Journal of Public Health, 34*(s1), S87–S92.

Dwyer, J., O'Donnell K, Lavoie J., Marlina U., & Sullivan P. (2009). *The overburden report: Contracting for Indigenous health services*. Flinders University and Cooperative Research Centre for Aboriginal Health. http://www.lowitja.org.au/sites/default/files/docs/Overburden%20summary%20brochure.pdf

Dwyer, J. M., Lavoie, J., O'Donnell, K., Marlina, U., & Sullivan, P. (2011). Contracting for Indigenous health care: Towards mutual accountability. *The Australian Journal of Public Administration, 70*(1), 34–46.

Ezzy, D. (2003). *Qualitative analysis*. Routledge.

Fogarty W., Bulloch H., McConnell S., & Davis M. (2018). *Deficit discourse and indigenous health: How narrative framings of Aboriginal and Torres Strait Islander people are reproduced in policy*. The Lowitja Institute and National Centre for Indigenous Studies, Canberra. https://www.lowitja.org.au/page/services/resources/Cultural-and-social-determinants/racism/Deficit-Discourse-and-Indigenous-Health

Foley, G. (1991). Redfern Aboriginal Medical Service: 20 Years On'. *Aboriginal and Islander Health Worker Journal, 15*(4), 4–8.

Foley, G. (2009). Black Power in Redfern 1968 – 1972. In K. de Souza & Z. Begg (Eds.), *There goes the neighbourhood: Redfern and the politics of urban space* (pp. 12–21). Break Down Press.

Ford, L. (2012). Locating indigenous self-determination in the margins of settler sovereignty: An introduction. In L. Ford & T. Rowse (Eds.), *Between Indigenous and settler governance*. Taylor and Francis Group.

Freeman, T., Baum, F., Lawless, A., Labonte, R., Sanders, D., Boffa, J., Edwards, T., & Javanparast, S. (2016). Case study of an aboriginal community-controlled health service in Australia: Universal, rights-based, publicly funded comprehensive primary health care in action. *Health and Human Rights Journal, 18*(2), 93–108.

Grant, M., Wronski, I., Murray, R. B., & Couzos S. (2008). Aboriginal health and history. In S. Couzos & R. Murray (Eds.), *Aboriginal primary health care: An evidence-based approach* (3rd edn), Oxford University Press.

Hall, S. E., Bulsara, C. E., Bulsara, M. K., Leahy, T. G., Culbong, M. R., Henrie, D., & Holman, C. D. (2007). Treatment patterns for cancer in Western Australia: Does being Indigenous make a difference? *Medical Journal of Australia, 181*(4), 191–194.

Henderson, G., Robson, C., Cox, L., Dukes, C., Tsey, K., & Haswell, M. (2007). Social and emotional wellbeing of Aboriginal and Torres Strait Islander people within the broader context of the social determinants of health. In I. Anderson, F. Baum, & M. Bentley (Eds.), *Beyond Bandaids: Exploring the underlying social determinants of Aboriginal health* (pp. 136–164). Cooperative Research Centre for Aboriginal Health.

Hood, C. (1991). A public management for all seasons?'. *Public Administration, 69*(1), 3–19.

Hunt, J. (2008). *Between a rock and a hard place: Self-determination, mainstreaming and Indigenous community governance*, In J. Hunt, D. Smith, S. Garling and W. Sanders (Eds), Contested Governance: Culture, power and institutions in Indigenous Australia, Research Monograph No. 29, Centre for Aboriginal Economic Policy Research: The Australian National University, available at: https://press.anu.edu.au/publications/series/centre-aboriginal-economic-policy-researchcaepr/contested-governance

Hurley, C., Baum, F., Johns, J., & Labonte, R. (2010). Comprehensive primary health care in Australia: Findings from a narrative review of the literature. *Australasian Medical Journal, 1*(2), 147–152.

Jan, S., Conaty, S., Hecker, R., Bartlett, M., Delaney, S., & Capon, T. (2004). An holistic economic evaluation of an Aboriginal community-controlled midwifery programme in Western Sydney. *Journal of Health Services Research and Policy, 9*(1), 14–21.

Kelaher, M. A., Ferdinand, A. S., & Paradies, Y. (2014). Experiencing racism in health care: The mental health impacts for Victorian Aboriginal communities. *Medical Journal of Australia, 201*(1), 44–47. https://doi.org/10.5694/mja13.10503

Labonte, R. (1994). Health promotion and empowerment: Reflections on professional practice. *Health Education Quarterly, 21*(2), 253–268.

Lardner, K. L. (2009). Understanding the impact of self-determination on communities in crisis. *International Journal of Indigenous Health, 5*(2), 88–101. https://doi.org/10.3138/ijih.v5i2.28984

Larkins, S. L., Geia, L. K., & Panaretto, K. S. (2006). Consultations in general practice and at an Aboriginal community controlled health service: Do they differ? *Rural and Remote Health, 6*(560), 1–12. www.rrh.org.au/journal/article/560

Lovett, R. (2014). A History of health services for Aboriginal and Torres Strait Islander people. In O. Best & B. Fredericks (Eds.), *Yatdjuligin: Aboriginal and Torres Strait Islander nursing and midwifery care* (pp. 31–50). Cambridge University Press.

Lowitja Institute. (2021). *Leadership and legacy through crises: keeping our mob safe.* Close the Gap Campaign Steering Committee for Indigenous Health Equality. https://humanrights.gov.au/our-work/aboriginal-and-torres-strait-islander-social-justice/publications/close-gap-2021, accessed 01/02/2022

Lowitja Institute. (2020). *We nurture our culture for our future, and our culture nurtures us.* Close the Gap Campaign Steering Committee for Indigenous Health Equality. https://humanrights.gov.au/our-work/aboriginal-and-torres-strait-islander-social-justice/publications/close-gap-2020?_ga=2.226248154.1140435180.1618470148-439551922.1613444445

Lowitja Institute. (n.d). Overcoming governance, regulatory and performance burdens in ACCHSs: timeline brochure. https://www.lowitja.org.au/content/Document/PDF/overburden-ke-timeline.pdf

Marmot, M. G. (2005). Social determinants of health inequalities. *The Lancet, 365*(9464), 1099–1104. https://doi.org/10.1016/S0140-6736(05)71146-6

Marmot, M. G. (2011). Social Determinants and the Health of Indigenous Australians'. *Aboriginal and Islander Health Worker Journal, 35*(3), 21–22. https://doi.org/10.5694/j.1326-5377.2011.tb03086.x

Martini, A., Marlina, U., Dwyer, J., Lavoie, J., O'Donnell, K., & Sullivan, P. (2011). *Aboriginal community controlled health service funding: Report to the sector.* The Lowitja Institute and Flinders University, Carlton South and Bedford Park. https://www.lowitja.org.au/lowitja-publishing/L008.

Meagher G., & Goodwin, S. (2015). Introduction: Capturing marketisation in Australian social policy. In G. Meagher & S. Goodwin (Eds.), *Markets, rights and power in Australian social policy.* Sydney University Press.

Moran M, Porter D., & Curth-Bibb, J. (2014). Funding Indigenous organisations: Improving governance performance through innovations in public finance management in remote Australia closing the gap clearinghouse, *Australian Institute of Health and Welfare*, available at: https://www.aihw.gov.au/getmedia/cc5909f4-869f-4a75-aed9-e170d1e0a5b8/ctgc-ip11.pdf.aspx?inline=true, accessed on 01 November 2020.

Moore, S. P., Green, A. C., Bray, F., Garvey, G., Coory, M., Martin, J., & Valery, P.C. (2014). Survival disparities in Australia: an analysis of patterns of care and comorbidities among indigenous and non-indigenous cancer patients. *BMC Cancer, 14*(517). https://doi.org/10.1186/1471-2407-14-517

Moran, M., Porter, D., & Curth-Bibb, J. (2016). The Impact of Funding Modalities on the Performance of Indigenous Organisations'. *Australian Journal of Public Administration, 75*(3), 359–372.

Murray M., Nelson, G., Poland, B., Maticka-Tyndale, E., & Ferris, L. (2004). Assumptions and values of community health psychology. *Journal of Health Psychology, 9*(2), 323–333. https://doi.org/10.1177/1359105304040897

Murphy, M. (2014). Self-determination as a collective capability: The case of indigenous peoples. *Journal of Human Development and Capabilities*, 15(4), 320–334. https://doi.org/10.1080/19452829.2013.878320

National Aboriginal Community Controlled Health Organisation. (2018). Annual Report 2017–2018. http://www.naccho.org.au/wp-content/uploads/NACCHO-Annual-Report-2017-18.pdf

National Aboriginal Community Controlled Health Organisation [NACCHO]. (2015). *Submission—Senate Standing Committees on Finance and Public Administration Commonwealth Indigenous Advancement Strategy tendering processes, Submission 70.* Commonwealth Indigenous Advancement Strategy Tendering Processes. http://www.aph.gov.au/Parliamentary_Business/Committees/Senate/Finance_and_Public_Administration/Commonwealth_Indigenous/Submissions

National Aboriginal Community Controlled Health Organisation [NACCHO]. (2011). *Constitution,* http://www.naccho.org.au/wpcontent/uploads/NACCHO-CONSTITUTION-Ratified-Ver-151111-for-ASIC-.pdf

Ngaanyatjarra Pitjantjatjara Yankunytjatjara Women's Council Aboriginal Corporation [NPY Women's Council]. (1999). *Malparara way: the strength of working together.* NPY Women's Council.

Osborne, K., Baum, F., & Brown, L. (2013). *What works? A review of actions addressing the social and economic determinants of Indigenous health', Issues Paper no. 7 produced for the Closing the Gap Clearinghouse.* Australian Institute of Health and Welfare. https://www.aihw.gov.au/reports/indigenous-australians/what-works-a-review-of-actions-addressing-the-social-and-economic-determinants-of-indigenous-health/contents/table-of-contents

Panaretto, K. S., Wenitong, M., Button, S., & Ring, I. T. (2014). Aboriginal community controlled health services: Leading the way in primary care'. *Medical Journal of Australia, 200*(11), 649–652.

Panaretto, K. S., Lee, H. M., Mitchell, M. R., Larkins, S. L., Manessis, V., Buettner, P. G., & Watson, D. (2005). Impact of a collaborative shared antenatal care program for urban Indigenous women: A prospective cohort study'. *Medical Journal of Australia, 182*(10), 514–519. https://doi.org/10.5694/j.1326-5377.2005.tb00017.x

Paradies, Y., Harris, R. and Anderson, I. (2008). The impact of racism on indigenous health in Australia and Aotearoa: Towards a research agenda' discussion paper series No. 4. Cooperative Research Centre for Aboriginal Health and Flinders University. https://www.lowitja.org.au/sites/default/files/docs/Racism-Report.pdf

Pearson, O., Schwartzkopff, K., Dawson, A., Hagger, C., Karagi, A., Davy, C., Brown, A., & Braunack-Mayer, A. (2020). Aboriginal community controlled health organisations address health equity through action on the social determinants of health of Aboriginal and Torres Strait Islander peoples in Australia. *BMC Public Health, 20*(1). https://doi.org/10.1186/s12889-020-09943-4

Pholi, K., Black, D., & Richards, C. (2009). Is 'Close the Gap' a useful approach to improving the health and wellbeing of Indigenous Australians? *Australian Review of Public Affairs, 9*(2), 1–13.

Pope, C., Ziebland, S., & Mays, N. (2007). *Analysing qualitative data.* IQualitative Research in Health Care. doi:https://doi.org/10.1002/9780470750841.ch7

Productivity Commission. (2017). *Introducing competition and informed user choice into human services: Identifying sectors for reform.* Commonwealth of Australia. https://www.pc.gov.au/inquiries/completed/human-services/identifying-reform/report

Rigney, L. I. (1999). Internationalization of an Indigenous anticolonial cultural critique of research methodologies: A guide to indigenist research methodology and its principles. *Wicazo Sa Review, 14*(2), 109–121.

Rowse T. (2005). The indigenous sector. In D. Austin-Broos & G. Macdonald (Eds.), *Culture, economy and governance in aboriginal Australia: Proceedings of a workshop of the academy of social sciences in Australia* (pp. 213–230). Sydney University Press.

Rowse, T. (2012). *Rethinking social justice: From 'peoples' to 'populations'*. Aboriginal Studies Press.

Russell, L. (2014). *Impact of the 2014–25 budget on indigenous programs and services*. Menzies Centre for Health Policy, University of Sydney. https://ses.library.usyd.edu.au/bitstream/2123/11442/1/2014-15IndigenousProvisions.pdf.

Schrecker, T., & Bambra, C. (2015). *How politics makes us sick: Neoliberal epidemics*. Palgrave Macmillan

Senate Finance and Public Administration References Committee [Senate FPARC]. (2016). *Report: Commonwealth Indigenous Advancement Strategy tendering processes*. Commonwealth of Australia. http://www.aph.gov.au/Parliamentary_Business/Committees/Senate/Finance_and_Public_Administration/Commonwealth_Indigenous/Report

Simpson, A. (2014). *Mohawk interruptus: Political life across the borders of settler states*. Duke University Press.

Strakosch, E. (2019). The technical is political: Settler colonialism and the Australian Indigenous policy system. *Australian Journal of Political Science, 54*(1), 114–130. https://doi.org/10.1080/10361146.2018.1555230

Taylor, K., & Guerin, P. (2010). *Health care and indigenous Australians: Cultural safety in practice*. Palgrave Macmillan.

Thomas, D. P., Heller, R. F., & Hunt, J. M. (1998). Clinical consultations in an aboriginal community-controlled health service: A comparison with general practice. *Australian and New Zealand Journal of Public Health, 22*(1), 86–91. https://doi.org/10.1111/j.1467-842X.1998.tb01150.x

Thornton, P. H., Ocasio, W., & Lounsbury, M. (2012). *The institutional logics perspective: A new approach to culture, structure and process*. Oxford University Press.

Tuhiwai-Smith, L. (1999). *Decolonizing methodologies: Research and Indigenous peoples*. Zed Books.

United Nations. (2007). *The United Nations declaration on the rights of Indigenous peoples*. https://www.un.org/development/desa/indigenouspeoples/wp-content/uploads/sites/19/2018/11/UNDRIP_E_web.pdf

van Ryan, M., & Fu, S. S. (2011). Paved with good intentions: Do public health and human service providers contribute to racial/ethnic disparities in health? *American Journal of Public Health, 93*(2), 248–255.

Wallerstein, N (2002). Empowerment to reduce public health disparities. *Scandinavian Journal of Public Health, 30*(59), 72–77.

Wilkinson, R. G., & Marmot, M. (Eds.). (2006). *The social determinants of health* (2nd edn.). Oxford University Press.

# Chapter 6
# Stepping Stones to Indigenous Futures: Rethinking Precarity in Indigenous Education and Work

**Nikki Moodie**

**Abstract** This chapter introduces the new education and employment targets in the latest iteration of the federal government's Closing The Gap (CTG) policy, and discusses these in the context of Aboriginal and Torres Strait Islander peoples' precarious engagements in school and work. Drawing on Tuck and Yang's (2012) work on incommensurability and Walter's (2022) sociology of Indigenous lifeworlds, this chapter contends that the 'stepping stones' approach to education and work embedded in current policy settings is not a path laid by Indigenous people towards Indigenous futures. For decades, government policy has maintained an ideological commitment to 'jobs pathways', despite mounting evidence that Indigenous people experience significant precarity in our education, training, and work careers. Despite recent changes to the Commonwealth Government's Closing the Gap policy—which does include increased recognition of Indigenous rights—this chapter argues that it is the absence of collectively-determined pathways that cements precarity. Indigenous peoples' learning and work pathways *should* be linear, not attenuated and partial or marked by racism and marginalisation. It is the absence of self-determination that compounds precarity, and whilst the refreshed CTG targets do include greater recognition of Indigenous lifeworlds and the right to determine our own futures and aspirations, the 'stepping stones' on offer are neither easy to walk nor lead to self-determined Indigenous futures.

**Keywords** Closing the gap · Policy · Settler colonialism · Transitions · Life course

## 6.1 Introduction

The analogy of 'stepping stones' through education and employment is a powerful idea that contains the promise of smooth transitions and linear trajectories through the life course, but which, in reality, is an idea whose time has well and truly passed (Hunter, 2010, p. 4; OECD, 2009, p. 60; te Riele, 2004). Whilst a positive relationship

N. Moodie (✉)
The University of Melbourne, Parkville, VIC, Australia
e-mail: nikki.moodie@unimelb.edu.au

 
N. Moodie and S. Maddison (eds.), *Public Policy and Indigenous Futures*,
Indigenous-Settler Relations in Australia and the World 4,
https://doi.org/10.1007/978-981-19-9319-0_6

between higher levels of education and more stable, higher paid employment does exist for Indigenous people, there remain significant differences in school and work participation according to gender, remoteness, and age (Venn, 2016; Young et al., 2007). Compounding these differences are variable labour market conditions (Venn, 2016), rising inequality within Indigenous communities (Markham & Biddle, 2018), and exposure to economic shocks as a result of the COVID-19 pandemic (Dinku et al., 2020). This evidence suggests that the idea that there are 'stepping stones' for Indigenous people to follow from full-time schooling to full-time work not only misrecognises 'the steady erosion of employment opportunities' for an entire generation of Australians (Wyn & Woodman, 2021 p. 466), but also the impact of racism and discrimination on Indigenous people's lives, especially at school and work (Biddle et al., 2013; Dinku et al., 2020; Moodie et al., 2019; Paradies & Cunningham, 2009). Education and labour policies that focus on individual skills and attributes, such as Year 12 completion or participation rates, prioritise personal 'labour market endowments' (Dinku et al., 2020, p. 191) as one of the primary mechanisms to alleviate Indigenous disadvantage. Such policies neglect the reality that 'stepping stones' have been displaced by deregulated labour markets and subsequent economic dislocation (Wyn & Woodman, 2021) as well as the specificities of Indigenous peoples' experiences of racism of school and work in Australia (Moodie et al., 2019). The Australian Government's flagship Indigenous affairs policy, Closing the Gap, not only represents the relationship between school and work as linear, but in so doing locates sole responsibility for precarity with the person, wholly avoiding structural explanations and curtailing the possibility of collectively defined Indigenous pathways to meaningful Indigenous lives.

## 6.2 Closing the Gap

Initially released in 2008, the Australian Government's Closing the Gap policy has for more than a decade defined the representation of Indigenous policy problems and solutions in a range of sectors (Dawson et al., 2020; Gardiner-Garden, 2013). Closing the Gap has since its inception been the subject of trenchant critique which has highlighted the policy's deficit framing of Indigenous agency, its assimilationist objectives, and persistent failure to achieve the aims it originally outlined (Altman & Fogarty, 2010; Dawson et al., 2020; Pholi et al., 2009). Despite the continued failure of CTG to show material improvement in the lives of Aboriginal and Torres Strait Islander peoples, and despite persistent and substantial critique, CTG was 'refreshed' and updated in 2020. Boasting a deeper engagement with Indigenous service delivery organisations and new targets, the extended CTG framework now aims to collect data and report progress on a much wider array of social indicators than the initial six targets (see Fig. 6.1).

The refreshed CTG policy includes four Priority Reform areas: partnership agreements for CTG implementation; increasing funding and government contracts to Indigenous service delivery organisations; decreasing the number of Indigenous

Initial Closing the Gap targets

1. to close the life expectancy gap within a generation
2. to halve the gap in mortality rates for Indigenous children under five within a decade
3. to ensure access to early childhood education for all Indigenous four year olds in remote communities within five years
4. to halve the gap in reading, writing and numeracy achievements for children within a decade
5. to halve the gap for Indigenous students in year 12 attainment rates by 2020 and
6. to halve the gap in employment outcomes between Indigenous and non-Indigenous Australians within a decade.

**Fig. 6.1** National Indigenous Reform Agreement 2008 original Closing the Gap targets

people who report racism in the public service; and, data sharing partnerships. The Priority Reform areas are supplemented by 17 Socioeconomic Outcome Areas (SOA), each with identified Targets and Supporting Indicators (see Table 6.1).[1]

The involvement of the Coalition of Aboriginal and Torres Strait Islander Peak Organisations (Coalition of Peaks) in the redevelopment of CTG and the inclusion of new indicators suggests that even though many of the deep-seated problems in this policy framing remain, there are small, partial shifts that may allow the possibility that Indigenous futures can be more fully recognised. Whilst it is beyond the scope of this chapter to interrogate the concordance between all of the new Outcomes and their designated Indicators or Measures (for example, noting the equation in SOA14 between wellbeing and suicide reduction, or in SOA17 between decision-making and digital inclusion), it is positive to note that most Indicators are not seeking to measure progress against the non-Indigenous population (Walter et al., 2019). Concerns do remain, however. It may be suggested, for example, that the Indicator for Socioeconomic Outcome Area 15—the 'area of Australian land mass and sea waters that is subject to Aboriginal and Torres Strait Islander people's legal rights or interests'—reinforces the limited position of Indigenous rights in Australia (Atkinson, 2001; Short, 2016). Noting that such rights are only incrementally won or recognised here (Anderson, 2015), the rewrite of federal policy to positively acknowledge the importance of Indigenous peoples' connection to Country brings CTG into closer alignment with a recognition discourse that seeks to incorporate Indigenous peoples into the liberal polity. Whether this reflects the aspirations of all Indigenous peoples' or collectives on the continent, or the interests of the stakeholder groups represented in the CTG refresh is an ongoing discussion. So whilst the Outcome Areas are operationalised in sometimes problematic ways, reflecting the nature of state interests and capacities in data definition, collection and sharing (Lovett et al., 2020), and the track record of this policy approach failing to achieve positive outcomes is well established (Bond & Singh, 2020), the inclusion of high level language related to

[1] See https://www.pc.gov.au/closing-the-gap-data/dashboard

**Table 6.1** 2020 Closing the Gap targets

| Socioeconomic outcome area | Target | Indicator |
|---|---|---|
| 1. Aboriginal and Torres Strait Islander people enjoy long and healthy lives | Close the Gap in life expectancy within a generation, by 2031 | Life expectancy |
| 2. Children are born healthy and strong | By 2031, increase the proportion of Aboriginal and Torres Strait Islander babies with a healthy birthweight to 91 per cent | The proportion of Aboriginal and Torres Strait Islander babies with a healthy birthweight |
| 3. Aboriginal and Torres Strait Islander children are engaged in high quality, culturally appropriate early childhood education in their early years | By 2025, increase the proportion of Aboriginal and Torres Strait Islander children enrolled in Year Before Fulltime Schooling (YBFS) early childhood education to 95 per cent | The proportion of Aboriginal and Torres Strait Islander children in the state specific YBFS age cohort who are enrolled in a preschool program |
| 4. Aboriginal and Torres Strait Islander children thrive in their early years | By 2031, increase the proportion of Aboriginal and Torres Strait Islander children assessed as developmentally on track in all five domains of the Australian Early Development Census (AEDC) to 55 per cent | The proportion of Aboriginal and Torres Strait Islander children assessed as developmentally on track in all five domains of the AEDC |
| 5. Aboriginal and Torres Strait Islander students achieve their full learning potential* | By 2031, increase the proportion of Aboriginal and Torres Strait Islander people (age 20–24) attaining year 12 or equivalent qualification to 96 per cent | The proportion of Aboriginal and Torres Strait Islander people aged 20–24 years who have attained a minimum of Year 12 or equivalent, or Certificate level III or above qualification |
| 6. Aboriginal and Torres Strait Islander students reach their full potential through further education pathways[a] | By 2031, increase the proportion of Aboriginal and Torres Strait Islander people aged 25–34 years who have completed a tertiary qualification (Certificate III and above) to 70 per cent | The proportion of Aboriginal and Torres Strait Islander people aged 25–34 years old who have completed qualifications at Certificate level III or above |
| 7. Aboriginal and Torres Strait Islander youth are engaged in employment or education* | By 2031, increase the proportion of Aboriginal and Torres Strait Islander youth (15–24 years) who are in employment, education, or training to 67 per cent | The proportion of Aboriginal and Torres Strait Islander youth aged 15–24 years who are fully engaged in employment, education, or training |
| 8. Strong economic participation and development of Aboriginal and Torres Strait Islander people and communities* | By 2031, increase the proportion of Aboriginal and Torres Strait Islander people aged 25–64 who are employed to 62 per cent | The proportion of Aboriginal and Torres Strait Islander people aged 25–64 years who are employed |
| 9. Aboriginal and Torres Strait Islander people secure appropriate, affordable housing that is aligned with their priorities and need* | By 2031, increase the proportion of Aboriginal and Torres Strait Islander people living in appropriately sized (not overcrowded) housing to 88 per cent | The proportion of Aboriginal and Torres Strait Islander people living in appropriately sized (not overcrowded) housing |

(continued)

**Table 6.1** (continued)

| Socioeconomic outcome area | Target | Indicator |
|---|---|---|
| 10. Aboriginal and Torres Strait Islander people are not overrepresented in the criminal justice system | By 2031, reduce the rate of Aboriginal and Torres Strait Islander adults held in incarceration by at least 15 per cent | Age-standardised imprisonment rate |
| 11. Aboriginal and Torres Strait Islander young people are not overrepresented in the criminal justice system | By 2031, reduce the rate of Aboriginal and Torres Strait Islander young people (10–17 years) in detention by at least 30 per cent | The rate of Aboriginal and Torres Strait Islander young people aged 10–17 years in detention |
| 12. Aboriginal and Torres Strait Islander children are not overrepresented in the child protection system | By 2031, reduce the rate of overrepresentation of Aboriginal and Torres Strait Islander children (0–17 years old) in out-of-home care by 45 per cent | The rate of Aboriginal and Torres Strait Islander children aged 0–17 years in out-of-home care |
| 13. Aboriginal and Torres Strait Islander families and households are safe | By 2031, the rate of all forms of family violence and abuse against Aboriginal and Torres Strait Islander women and children is reduced at least by 50 per cent, as progress towards zero | Proportion of Aboriginal and Torres Strait Islander females aged 15 years and over who experienced domestic physical or threatened physical harm in the last 12 months |
| 14. Aboriginal and Torres Strait Islander people enjoy high levels of social and emotional wellbeing | A significant and sustained reduction in suicide of Aboriginal and Torres Strait Islander people towards zero | Suicide death rate of Aboriginal and Torres Strait Islander people |
| 15. Aboriginal and Torres Strait Islander people maintain a distinctive cultural, spiritual, physical, and economic relationship with their land and waters | By 2030, a 15 per cent increase in: Australia's land mass subject to Aboriginal and Torres Strait Islander people's legal rights or interests (Target 15a) Australia's sea waters subject to Aboriginal and Torres Strait Islander people's legal rights or interests (Target 15b) | Area of Australian land mass and sea waters that is subject to Aboriginal and Torres Strait Islander people's legal rights or interests |
| 16. Aboriginal and Torres Strait Islander cultures and languages are strong, supported, and flourishing | By 2031, there is a sustained increase in number and strength of Aboriginal and Torres Strait Islander languages being spoken | The number and strength of Aboriginal and Torres Strait Islander languages being spoken |
| 17. Aboriginal and Torres Strait Islander people have access to information and services enabling participation in informed decision-making regarding their own lives | By 2026, Aboriginal and Torres Strait Islander people have equal levels of digital inclusion | Proportion of Aboriginal and Torres Strait Islander people aged 15 years and over who have accessed the internet at home in the last 12 months |

[a]No baseline data available since 2016

land and language does send a signal that Indigenous futures are dependent on these interconnected aspects of Indigenous life (Finlay et al., 2021).

Closing the Gap has always made a claim on the future, explicitly through setting dates by which rates of disadvantage should be halved, for example, or implicitly through its representation of what a good and healthy Indigenous life should look like. In the first iteration of CTG, the Aboriginal or Torres Strait Islander person it sought to imagine was statistically identical to a non-Indigenous neoliberal subject, with all the assimilationist objectives that implies (Altman & Fogarty, 2010; Pholi et al., 2009). In the latest iteration of CTG, the Aboriginal and Torres Strait Islander people it seeks to imagine might speak their languages or have their rights to and interests in land recognised. But the pathways to those futures are not identifiably characterised by Indigenous institutions or defined by Indigenous governance in ways that offer an alternative to participation in an economic system that has long lost its ability to provide meaningful transitions for young people. The language of 'participation' and 'potential' in the refreshed CTG still implies that the primary barriers we face are of our own making and that attendance at school and full-time work—as in SOA5, 6, 7 and 8—will ensure easier and more stable transitions through the Indigenous life course. This misrecognises the nature of Indigenous peoples' experiences of schooling and the labour market, and in this rendering, Indigenous peoples' self-determined pathways through our own understandings of what 'learning' and 'work' mean become impossible imaginings. Stepping stones, as illusory or attenuated as they may be in liquid modernity, remain a powerful analogy for a life course that has purpose, meaning, and predictability. When the settler state imposes learning and work pathways, such as those sedimented in the CTG framework, those stepping stones lead anywhere other than Indigenous futures.

## 6.3 Precarity in Work and Education

Providing an analysis of contemporary labour force issues illustrates the non-linear features of Indigenous employment that, whilst describing precarity, also illustrates a more persistent set of practices that deploy settler colonial notions of class, race, and gender in an attempt to undermine bases of attachment and meaning-making for Indigenous people. Whilst the new CTG framework offers some acknowledgement of what makes a meaningful Indigenous life through language and connection to Country, the pathway to get there remains limited to the same forms of participation that previous iterations of the policy offered.

As discussed above, the stepping stone assumption that underpins much policy and research in the field of education, training, and employment—including in state-driven Indigenous policy—suggests that holding any job increases a person's chances of finding better paying, more satisfying, or more stable employment. Thus, 'stepping stones' imply that a bad job will eventually lead to a good job, and so, any job is better than none (OECD, 2009, pp. 59–60). This assumption has not only been disrupted by the economic fallout of the COVID-19 pandemic, particularly regarding negative participation rates for young men and rising underutilisation rates for young women (Churchill 2021), but by even larger social changes that describe adulthood itself increasingly in terms of 'precariousness and insecurity' (Wyn & Woodman, 2021). There is further evidence that the stepping stone assumption does not work the same way for Indigenous Australians, for whom insecure work already predicts an exit from the workforce (Hunter & Gray, 2017). This precarity is exacerbated for young Indigenous women (aged 20 to 29), who are less likely to enter employment, more likely to enter the workforce on insecure and part-time contracts, and more likely to become unemployed than other Australians. Overall, Indigenous people are more likely to be marginally attached and discouraged workers—wanting work but given up looking—due to a lack of job opportunities and racism and discrimination in the workplace (Hunter, 2010; Hunter & Gray, 2017, p. 8).

Analysing Indigenous employment trends from 1994 to 2014, Gray and Hunter (2016) consider the impact of an extended national period of strong economic growth largely driven by expansion of extractive industries, and the subsequent economic retraction following the Global Financial Crisis in 2009. This period saw an increase in Indigenous employment by 20 per cent and a reduction in the disparity between Indigenous and non-Indigenous employment outcomes (Gray & Hunter, 2016, p. 1). Nonetheless, the initial rapid growth in employment rates between 1994 and 2008 has not been sustained, with Indigenous men experiencing greater levels of marginal attachment to the labour force in the period to 2014. Between 1994 and 2008, the employment rate for Indigenous men grew from 38 per cent to 59 percent, but then plateaued (Gray & Hunter, 2016, p.3). The number of unemployed Indigenous men has fallen over the same time, but this has not been driven by increasing employment. Rather, this signals that Indigenous men were exiting the labour force and becoming marginally attached, wanting work but unable to find it. The marginal attachment rate increased from 8 per cent to 13 per over the 2008–2015 period and explains the overall fall in the Indigenous male unemployment rate (Gray & Hunter, 2016, p.3). Similarly, the employment rate for Indigenous women increased significantly from 25 per cent to 43 per cent over the 1994 to 2008 period, but then almost no change to 2014. There was no change in the proportion of Indigenous women who wanted to work and couldn't find it, and only a small increase of 2 per cent in the proportion of Indigenous women who were out of the labour force (Gray & Hunter, 2016, p. 2). The authors conclude that it was not a lack of willingness to participate in the labour force, but rather changes in the nature of work and its availability, that accounted for these fluctuations (Gray & Hunter, 2016, p. 5).

In considering non-linear pathways from education and training to employment as a useful strategy to understand Indigenous peoples' experiences, the following points are of note:

- An Indigenous man or woman with a degree has an employment probability of 85 per cent or 74 per cent respectively. For someone who has completed Year 12 only, this falls to 62 per cent and 50 per cent, whereas someone who has completed Year 9 or below it falls to 43 per cent and 32 per cent (Biddle et al., 2016).

- Although education outcomes overall and Year 12 attainment rates in particular have improved for Indigenous young people (Australian Bureau of Statistics (ABS) 2017), with some 66 per cent of Indigenous people aged 20 to 24 completing Year 12 or equivalent, these rates vary according to geographic location (Commonwealth of Australia, 2020). Given the younger demographic and variability in completion rates, a 'business as usual' approach to education and training is unlikely to respond to the learning needs of people on Country, or be able to address any gendered differences in completion rates (Guenther & McRae-Williams, 2014).
- Young Indigenous women who do not complete Year 12 are likely to spend around 10 years not studying before enrolling in post-secondary studies. Often family responsibilities, raising young families, or cultural obligations account for differing patterns of investment in education (Hunter & Gray, 2017).
- Indigenous people are 1.4 times more likely than non-Indigenous Australians to start a university qualification over the age of 45 (Australian Health Ministers' Advisory Council, 2017).
- The first Aboriginal people to graduate with tertiary qualifications did so in 1959 and 1966; Margaret Williams-Weir and Charles Perkins. Two generations later, there are 22,935 Indigenous people attending a tertiary institution nationwide (Department of Education, Skills and Employment, 2022). Despite this rapid increase, it is important to note that family and friends are likely to remain the most frequently consulted sources for career advice for both Indigenous and non-Indigenous young people (Craven et al., 2005, p. 17). Given historical exclusion from formal education and training, and increasing but still low tertiary entrance rates, many Indigenous people seek post-school learning or employment opportunities without the benefit of accumulated intergenerational knowledge regarding access, expectations, navigation, or outcomes (Fredericks et al., 2015; Gore et al., 2017).

Li (2014) considers labour market outcomes related to level of education for Indigenous people with university qualifications. Recognising that 4.9 per cent of Aboriginal and Torres Strait Islander people have a bachelor's degree or higher (ABS, 2016), compared to 31.4 per cent (ABS, 2018) of the wider population, increasing the prevalence of higher education is a key aim of national policy positions including the 'refreshed' CTG (Commonwealth of Australia, 2020ustralia 2020; Universities Australia, 2017). Of course, Indigenous people with post-school qualifications are more likely be employed than those without (Australian Institute of Health & Welfare, 2020). Further, Li's (2014) analysis noted that Indigenous people with higher education qualifications are less likely to receive a financial penalty for being overeducated, that is, less likely to be employed in jobs below their level of education. However, Indigenous graduates earn on average 3% less than their non-Indigenous counterparts (Li, 2014, p.3). Not only is this exacerbated for Indigenous women graduates who earn 8% less than Indigenous men, but for those Indigenous graduates in the bottom decile of earners, the wage gap increases to a staggering 27% when compared to non-Indigenous graduates in the same position (Li, 2014).

Pathways to education, training and employment for Indigenous people do not have the same characteristics as pathways for non-Indigenous and other Australians. In part this is due to supply-side factors that attenuate Indigenous achievement in the early years; access to pre-school, culturally exclusive curriculum and testing regimes, lower expectations from teachers and schools, and discrimination, combine to ensure that Indigenous peoples' aspirations are over time adjusted downwards in line with social opportunities (Biddle & Cameron, 2012, p.7; Moodie et al., 2019). But also missing from these analyses is the recognition that Indigenous lifeworlds contain sovereign struggles and aspirations; desires and hopes for Indigenous futures that might well include participation in settler economies, but which may also contain the revitalisation of our own socio-political and economic ordering (Finlay et al., 2021; Guenther & McRae-Williams, 2014; Vizenor, 2008). As Walter describes:

> As Indigenous Peoples we retain our thousands of years of deep history of our lands, culture, traditions, and ways of knowing and being. These distinguish and shape our lived realities and the meanings embedded in their associated epistemologies, narratives, and logics also frame the Indigenous life world ... Any conceptualization of the Indigenous lifeworld must incorporate specific Indigenous social and cultural life circumstances. (Walter, 2022 p. 8)

A higher level of education is usually associated with improved health and wellbeing (Australian Institute of Health and Welfare, 2020). Some research has troubled this claim in Indigenous contexts, with the ABS suggesting that a higher level of education does not necessarily lead to a reduction in health risk behaviours (ABS, 2011) and the importance of the bi-directional relationships between health and education often overlooked (Anderson et al., 2017). Moreover, whilst the direct effect of level of education on employment is significant, the relationship is heavily dependent on the degree of empowerment, measured as identity, self-efficacy and resilience (Wilson et al., 2019). As Wilson et al. (2019) describe:

> ...successful pathways between education and employment are contingent upon their capacity to build from empowerment. These findings suggest that the ability of policy and practice in education and employment to improve outcomes depends on their capacity to genuinely empower people. That is the more programs empower people the more likely they are to succeed. (Wilson et al., 2019, p. 4).

Finlay et al. (2021) concur. In their analysis of the cultural determinants of health, Finlay et al. note that 'culture is significantly and positively associated with physical health, social and emotional wellbeing, and reductions in behaviour that is detrimental to health and wellbeing' (p. S13). So, whilst increasing education rates has a strong effect on employment (Thapa et al., 2012) and may even close most of the gap in employment levels (ABS, 2014), the *degree* of empowerment underlies achievement in education (Wilson et al., 2019, p. 5). Any direct relationship between education and employment is heavily mediated by the degree to which Indigenous people feel strong about our identity, able to make changes in our lives when we need to, and feel able to move on from bad experiences (Wilson et al., 2019). Where language reclamation is seen as an integral part of learning and education, for example, wellbeing improves (Finlay et al., 2021, p. S13). Where children experience high levels of cultural continuity, they experience less racism and enjoy a lower burden of disease and chronic stress in adulthood (Currie et al., 2019). This suggests that where Indigenous people are able to lay our own stepping stones to our own futures, we are healthier, happier, more connected to our families; self-determining our own learning pathways for futures we decide are meaningful.

The changes to the Closing the Gap approach in Australia may signal a move towards recognising Indigenous values and life courses, but experiences of racism and discrimination continue to diminish Indigenous peoples' lives and opportunities, and the right to a life on Country is barely acknowledged. Based on an analysis of the 2008 National Aboriginal and Torres Strait Islander Social Survey (NATSISS), Dockery (2012 p. 290) observes that a stronger cultural identity is associated with higher educational attainment and employment rates as well as better mental health and feelings of happiness. However, there was some evidence to suggest that a strong cultural identity led to increased psychological stress for Indigenous people living in non-remote areas. Dockery argued that there may be a penalty to pay for maintaining a strong sense of cultural identity; psychological stress as a result of discrimination (Dockery, 2012, p. 299). So, whilst Indigenous peoples' wellbeing is improved in line with the strength of cultural connection, the maintenance of that identity in contexts that undermine, diminish, or neglect Indigenous cultures may reduce wellbeing. As Dockery suggests:

> … a strong sense of self-identity is not the same thing as a strong sense of persistence of that self-identity through time – those who do strongly identify with Indigenous culture may also suffer psychologically from doubts over the survival of that culture, and what their role would be should their connection with that culture be severed. (Dockery, 2012, p. 294)

Results from the Reconciliation Barometer suggest that Indigenous people feel culturally empowered 'only sometimes' in settler colonial institutional contexts, and that there has been little shift in the number of Indigenous Australians who feel like things will get better for them—at home, at work, or regarding their financial situation—across the 2014, 2016, 2018 and 2020 reports (Reconciliation Australia, 2020). More than half of respondents in the Barometer sample indicated they could either never or only sometimes be true to their culture at work, in education settings, or in interactions with government. This rises to 59 per cent when considering interactions with the legal system (Reconciliation Australia, 2020, p. 118). Since 2014, Aboriginal and Torres Strait Islander people report increasing levels of racial discrimination in all professional and service contexts; 'Most notably, there have been sharp increases regarding racial discrimination by police (17%), and by employers (17%), compared to six years ago' (Reconciliation Australia, 2020, p. 116).

A consideration of transgenerational experiences of racial discrimination and the impact of racism at school and work on transitions in the life course, cannot be made in isolation from an acknowledgement of the recency of recognition discourse (Corntassel, 2008). The long association between skin colour, behaviour, and class status has marked the assimilationist drive of Australia's political, religious, and economic institutions (Carlson, 2016), and the desire of settler colonial institutions for a white and reconciled future. Labour and class distinctions have often been imposed on

Indigenous people with the intent of codifying a reduction in the degree of indigeneity and thus claims to cultural continuity and Country. A consideration of education 'pathways' for Indigenous people cannot be separated from the assimilationist violence that has characterised our experience of school and work for generations.

> Discussion of who and what counts as being Aboriginal occurs within the context of attempts to understand impediments to assimilation or economic participation in the wider society. This is measured along a colour/culture continuum, based on degrees of blood admixture and degrees of distance from white society. Understandings of Aboriginal people are firmly co-opted into European social theories of class via established analytical research frameworks. Concepts of class—concepts entirely alien to Aboriginal societies—are introduced in relation to behaviour to differentiate, relate and understand Aboriginal subgroups along 'diluting' colour lines. These studies reflect and also support the thinking that overcoming the disadvantages and negative aspects of being Aboriginal requires leaving the Aboriginal ethos behind, and that this is easier to do as colour lightens. (Carlson, 2016, p. 31)

Unsurprisingly, supposed markers of middle-class affluence or security do not tend to ameliorate conditions of ontological insecurity for Indigenous people, processes of healing and reconnection following intergenerational trauma, or compromise cultural obligations and expectations (Carlson, 2016; Dockery, 2012; Lahn, 2013). As Carlson notes though, 'Aboriginal inversion of the signifiers of colonial classifications' (2016, p. 31) has a long history. The role of ideas like 'learning' and 'work' in the construction of contemporary Indigenous identities has been limited by the state and compounded by contemporary precarity. Yet, when learning is culturally grounded, progresses to more advanced skills and knowledges, and is understood as a relational and generative activity involving Country, kin and wellbeing; expertise is accumulated and recognised, and demonstrates precisely the stepping stones and linear transitions that create and maintain Indigenous social norms.

## 6.4 Conclusion

To focus solely on the contemporary insecurity of Indigenous workers and employment trends misrecognises the ways in which the settler colonial state has relied on indentured servitude, slavery, and wage theft to extract wealth from Indigenous bodies and lands; the gendered dimensions of which have enabled Australia's White patriarchal agrarian mythology. New targets in the CTG policy and the greater involvement of the Indigenous service delivery sector in the development of those targets, may indicate a deepening recognition of the ways in which Indigenous lifeways are different to the lives of White and non-Indigenous people of colour in Australia. Particularly with regard to engagement in education and employment, greater recognition of Indigenous languages and connection to Country may allow deeper conversations about what it means to pursue an Indigenous future in the Australian settler state. Whilst there has been a move away from comparing Indigenous people to non-Indigenous people in the newest iteration of CTG, and characteristics of the non-Indigenous population are not necessarily established as parity goals, this policy still has some distance to travel before it is able to accurately represent the challenges that Indigenous people face, and the aspirations that we hold.

## References


Anderson, I. (2015). The crisis of Australia's indigenous policy. *Meanjin, 74*(3), 54–59. https://doi.org/10.3316/informit.620556605198824

Anderson I., Lyons J. G., Luke J. N., & Reich H. S. (2017). Health determinants and educational outcomes for indigenous children. In M. Walter, K. Martin, & G. Bodkin-Andrews (Eds.), *Indigenous children growing up strong*. Palgrave Macmillan. https://doi.org/10.1057/978-1-137-53435-4_12

Atkinson, W. (2001). 'Not one iota' of land justice: Reflections on the Yorta Yorta native title claim 1994–2001. *Indigenous Law Bulletin, 5*(6), 19–23. http://classic.austlii.edu.au/au/journals/IndigLawB/2001/12.html

Australian Bureau of Statistics. (2011). *Education and indigenous wellbeing: Australian social trends*, March 2011, cat. no. 4102.0. https://www.abs.gov.au/AUSSTATS/abs@.nsf/Lookup/4102.0Main+Features50Mar+2011

Australian Bureau of Statistics. (2014). *Exploring the gap in labour market outcomes for Aboriginal and Torres Strait Islander People. Australian Social Trends*, 2014, cat. no. 4102.0. https://www.abs.gov.au/ausstats/abs@.nsf/Lookup/4102.0main+features72014#EQUAL

Australian Bureau of Statistics. (2016). *Education: National Aboriginal and Torres Strait Islander social survey*, 2014–15, cat. no. 4714.0. https://www.abs.gov.au/ausstats/abs@.nsf/Lookup/by%20Subject/4714.0~2014-15~Main%20Features~Education~5

Australian Bureau of Statistics. (2017, October 23). *Strong improvements in Aboriginal and Torres Strait Islander education outcomes*. http://www.abs.gov.au/ausstats/abs@.nsf/MediaRealesesByCatalogue/3DBA1B3973A0286ACA258148000D0DB8?OpenDocument

Australian Bureau of Statistics. (2018). *Education and work, Australia*, May 2018, cat. no. 6227.0. https://www.abs.gov.au/AUSSTATS/abs@.nsf/Lookup/6227.0Main+Features1May%202018?OpenDocument

Australian Health Ministers' Advisory Council. (2017). *Aboriginal and Torres Strait Islander health performance framework 2017 Report*. AHMAC. https://www.pmc.gov.au/sites/default/files/publications/indigenous/hpf-2017/tier2/206.html

Australian Institute of Health and Welfare. (2020). *Aboriginal and Torres Strait Islander Health Performance Framework 2020 summary report*. Cat. no. IHPF 2.AIHW. https://www.indigenoushpf.gov.au/

Altman, J., & Fogarty, B. (2010). Indigenous Australians as 'no gaps' subjects: Education and development in remote Australia. In I. Snyder & J. Nieuwenhuysen (Eds.), *Closing the gap in education? Improving outcomes in southern world societies* (pp. 110–128). Monash University Publishing.

Biddle, N., & Cameron, T. (2012). *Potential factors influencing Indigenous education participation and achievement*. NCVER. https://files.eric.ed.gov/fulltext/ED538263.pdf

Biddle, N., Howlett, M., Hunter, B., & Paradies, Y. (2013). Labour market and other discrimination facing Indigenous Australian. *Australian Journal of Labour Economics, 16*(1), 91–113. https://doi.org/10.3316/ielapa.363028735038386

Biddle, N., Hunter, B.H., Yap, M., & Gray, M. (2016, June 6). Eight ways we can improve Indigenous employment. *The Conversation*. https://theconversation.com/eight-ways-we-can-improve-indigenous-employment-60377

Bond, C. J., & Singh, D. (2020). More than a refresh required for closing the gap of Indigenous health inequality. *Medical Journal of Australia, 212*(5), 198-199.e1. https://doi.org/10.5694/mja2.50498

Carlson, B. (2016). *The politics of identity: Who counts as Aboriginal today?* Aboriginal Studies Press.

Craven, R., Tucker, A., Munns, G., Hinkley, J., Marsh, H., & Simpson, K. (2005). *Indigenous students' aspirations: dreams, perceptions and realities*. Department of Education, Science and Training, Commonwealth of Australia.

Churchill, B. (2021). COVID-19 and the immediate impact on young people and employment in Australia: A gendered analysis. *Gender Work Organ, 28*, 783–794. https://doi.org/10.1111/gwao.12563

Commonwealth of Australia. (2020). *Closing the gap report 2020*. Department of the Prime Minister and Cabinet. https://ctgreport.niaa.gov.au/

Corntassel, J. (2008). Toward sustainable self-determination: Rethinking the contemporary indigenous-rights discourse. *Alternatives, 33*(1), 105–132. https://doi.org/10.1177/030437540803300106

Currie, C. L., Copeland, J. L., & Metz, G. A. (2019). Childhood racial discrimination and adult allostatic load: The role of Indigenous cultural continuity in allostatic resiliency. *Social Science and Medicine, 241*, 112564. https://doi.org/10.1016/j.socscimed.2019.112564

Dawson, J., Augoustinos, M., Sjoberg, D., Canuto, K., Glover, K., & Rumbold, A. (2020). Closing the Gap: Examining how the problem of Aboriginal and Torres Strait Islander disadvantage is represented in policy. *Australia Journal of Social Issues, 1–17*, 105. https://doi.org/10.1002/ajs4.125

Department of Education, Skills and Employment. (2022). *Higher education statistics: 2020 Section 6 indigenous students*. https://www.dese.gov.au/higher-education-statistics/resources/2020-section-6-indigenous-students

Dinku, Y., Hunter, B., & Markham, F. (2020). How might COVID-19 affect the Indigenous labour market? *Australian Journal of Labour Economics, 23*(2), 189–210. https://doi.org/10.3316/informit.469159601936670

Dockery, A.M. (2012). Do traditional culture and identity promote the wellbeing of Indigenous Australians? Evidence from the 2008 NATSISS. In B. Hunter & N. Biddle (Eds.), *Survey analysis for indigenous policy in Australia: Social sciences perspectives* (pp. 281–306). ANU Press. https://www.jstor.org/stable/j.ctt24h8br

Finlay, S., Canuto, K., Canuto, K., Neal, N., & Lovett, R. W. (2021). Aboriginal and Torres Strait Islander connection to culture: Building stronger individual and collective wellbeing. *Medical Journal of Australia, 214*(8), S12-16.

Fredericks, B., Lamey, T., Mikecz, M., & Santamaria, F. (2015). Enabling people to "see what they can be": The Community Aspirations Program (CAP-ED). *Learning Communities: International Journal of Learning in Social Contexts, 17*, 54–63. https://doi.org/10.18793/LCJ2015.17.05

Gardiner-Garden, J. (2013). Closing the gap. In *Parliamentary library briefing book: Key Issues for the 44th Parliament*. Commonwealth of Australia. https://www.aph.gov.au/About_Parliament/Parliamentary_Departments/Parliamentary_Library/pubs/BriefingBook44p/ClosingGap

Gore, J., Patfield, S., Fray, L., Holmes, K., Gruppetta, M., Lloyd, A., Smith, M., & Heath, T. (2017). The participation of Australian Indigenous students in higher education: A scoping review of empirical research, 2000–2016. *The Australian Educational Researcher, 44*, 323–355. https://doi.org/10.1007/s13384-017-0236-9

Guenther, J., & McRae-Williams, E. (2014). *Does education and training for remote Aboriginal and Torres Strait Islanders lead to "real" jobs? Evidence from the 2011 Census*. Paper presented at the 17th Australian vocational education and training research association conference, Surfers Paradise, Queensland. http://hdl.voced.edu.au/10707/313058

Hunter, B. H. (2010). *Pathways for Indigenous school leavers to undertake training or gain employment* (Resource Sheet No. 2). Closing the Gap Clearinghouse and the Australian Institute of Health and Welfare. https://www.aihw.gov.au/reports/indigenous-australians/pathways-for-indigenous-school-leavers-to-undertak/contents/table-of-contents

Gray, M., & Hunter, B. H. (2016). *Indigenous employment after the boom* (CAEPR Topical Issue No. 1/2016). Centre for Aboriginal Economic Policy Research, Australian National Universitty.

Hunter, B. H., & Gray, M. (2017). Dynamics of indigenous and non-indigenous labour markets. *Australian Journal of Labour Economics, 20*(1), 1–20. https://businesslaw.curtin.edu.au/wp-content/uploads/sites/5/2018/05/AJLE-v20n1-Hunter-Gray.pdf
Lahn. J. (2013). *Aboriginal professionals: Work, class and culture* (Working Paper No. 89/2013). Centre for Aboriginal Economic Policy Research, Australian National University. http://caepr.cass.anu.edu.au/research/publications/aboriginal-professionals-work-class-and-culture
Li, I. W. (2014). Labour market performance of indigenous university graduates in Australia: An ORU perspective. *Australian Journal of Labour Economics, 17*(2), 87–110. Retrieved from: https://businesslaw.curtin.edu.au/wp-content/uploads/sites/5/2016/04/AJLE-v17n2-II.pdf
Lovett, R., Jones, R., and Maher, R. (2020). The intersection of indigenous data sovereignty and closing the gap policy in Australia. In M. Walter, T. Kukutai, S. Russo Carroll, & D. Rodriguez-Lonebear (Eds.), *Indigenous data sovereignty and policy*. Routledge. https://doi.org/10.4324/9780429273957
Markham, F. and Biddle, N. (2018). *Income, poverty and inequality* (CAEPR Census Paper No. 2). Centre for Aboriginal Economic Policy Research, Australian National University. https://openresearch-repository.anu.edu.au/bitstream/1885/145053/1/CAEPR_Census_Paper_2.pdf
Moodie, N., Maxwell, J., & Rudolph, S. (2019). The impact of racism on the schooling experiences of Aboriginal and Torres Strait Islander students: A systematic review. *The Australian Educational Researcher, 46*(2), 273–295. https://doi.org/10.1007/s13384-019-00312-8
Organisation for Economic Co-operation and Development. (2009). Jobs for Youth/Des emplois pour les jeunes: Australia. *OECD Publishing*. https://doi.org/10.1787/9789264060203-en
Paradies, Y., & Cunningham, J. (2009). Experiences of racism among urban Indigenous Australians: Findings from the DRUID study. *Ethnic and Racial Studies, 32*(3), 548–573. https://doi.org/10.1080/01419870802065234
Pholi, K., Black, D., & Richards, C. (2009). Is 'Close the Gap' a useful approach to improving the health and wellbeing of Indigenous Australians? *Australian Review of Public Affairs, 9*(2), 1–14.
Reconciliation Australia. (2018). *2018 Australian reconciliation barometer*. https://www.reconciliation.org.au/wp-content/uploads/2019/02/final_full_arb-full-report-2018.pdf
Reconciliation Australia. (2020). *2020 Australian reconciliation barometer*. https://www.reconciliation.org.au/publication/australian-reconciliation-barometer-2020/
Short, D. (2016). *Reconciliation and colonial power: Indigenous rights in Australia*. Taylor and Francis.
te Riele, K. (2004). Youth transition in Australia: Challenging assumptions of linearity and choice. *Journal of Youth Studies, 7*(3), 243–257. https://doi.org/10.1080/1367626042000268908
Thapa, P., Shah, Q., & Ahmad, S. (2012). What are the factors determining Indigenous labour market outcomes? In Hunter B. and Biddle N. (Eds.), *Survey analysis for indigenous policy in Australia: Social sciences perspectives* (pp. 125–162). ANU Press. http://www.jstor.org/stable/j.ctt24h8br.15
Tuck, E., & Yang, K. W. (2012). Decolonization is not a metaphor. *Decolonization: Indigeneity, Education and Society, 1*(1), 1–40.
Universities Australia. (2017*). Indigenous Strategy 2017–2020*. https://www.universitiesaustralia.edu.au/wp-content/uploads/2019/06/Indigenous-Strategy-v16-1.pdf
Venn, D. (2016). *Indigenous youth employment and the school-to-work transition* (2016 Census Paper No. 7). Centre for Aboriginal Economic Policy Research, ANU.
Vizenor, G. (2008). Aesthetics of survivance: Literary theory and practice. In G. Vizenor (Ed.), *Survivance: Narratives of native presence* (pp. 1–23). University of Nebraska Press.
Young, M., Guenther, J. and Boyle, A. (2007). *Growing the desert: educational pathways for remote Indigenous people*. NCVER. https://www.ncver.edu.au/research-and-statistics/publications/all-publications/growing-the-desert-educational-pathways-for-remote-indigenous-people
Walter, M., Banks, E., Lovett, R., & Thurber, K. (2019, 15 February). 'Refreshed' Close the Gap targets will focus on progress and achievement. *NITV*. https://www.sbs.com.au/nitv/article/2019/02/14/refreshed-close-gap-targets-will-focus-progress-and-achievement1

Walter, M. (2022). Conceptualizing and theorizing the indigenous lifeworld. In M. Walter, T. Kukutai, A. Gonzales, & R. Henry (Eds.), *Oxford handbook of indigenous sociology* (pp. 1–20). Oxford University Press. https://doi.org/10.1093/oxfordhb/9780197528778.013.1

Wilson, B., Abbott, T., Quinn, S., Guenther, J., McRae-Williams, E., & Cairney, S. (2019). Empowerment is the basis for improving education and employment outcomes for aboriginal people in remote Australia. *The Australian Journal of Indigenous Education, 48*(2), 153–161. https://doi.org/10.1017/jie.2018.2

Wyn, J., & Woodman, D. (2021). A New Adulthood: Conceptualizing Transition Using a Generational Framework. In E. A. Marshall and J. E. Symonds (Eds.), *Young adult development at the school-to-work transition: international pathways and processes*. Oxford University Press. https://doi.org/10.1093/oso/9780190941512.001.0001/oso-9780190941512-chapter-21.

# Chapter 7
# Treaty as a Pathway to Indigenous Controlled Policy: Making Space, Partnering, and Honouring New Relationships

**Sarah Maddison and Anya Thomas**

**Abstract** As several Australian jurisdictions embark on Australia's first treaty processes there is growing recognition of the extent to which treaty will recast Indigenous-state relations. The negotiation of treaties means the recognition of other sovereign authorities—not authorities to be created (as these have existed for millennia) but authorities that will require space to be exercised alongside the state. Bureaucracies that have understood their role as primarily one of service delivery to First Nations will have to reorient themselves to become treaty partners with First Nations seeking to exercise greater control and autonomy. While we cannot yet predict the outcome of these negotiations, nor is it appropriate for us to attempt to articulate First Nations' priorities, it is likely that, over time, treatied First Nations will seek to rewrite the policy relationship with government, pursuing autonomy and self-governance in the place of state authority and control. This chapter explores the possibilities and challenges of transforming public policy-making through treaty, arguing that it will take time to re-write the partnership manual and enable genuinely Indigenous-controlled policy to become the new political norm.

**Keywords** Treaty · Indigenous settler relations · Australia · Public administration · Indigenous peoples · First Nations

## 7.1 Introduction

After many decades of Indigenous advocacy, the negotiation of treaties between First Nations and the settler state is at last becoming a reality. Four Australian jurisdictions—Victoria, Queensland, Tasmania and the Northern Territory—have now committed to treaty processes with First Nations, and while these processes are still in their early days (with Victoria the most advanced), there is a growing realisation of the profound impact that treaty will have on public policy-making. State

S. Maddison (✉) · A. Thomas
The Australian Centre, University of Melbourne, Parkville, VIC, Australia
e-mail: sarah.maddison@unimelb.edu.au

 
N. Moodie and S. Maddison (eds.), *Public Policy and Indigenous Futures*, Indigenous-Settler Relations in Australia and the World 4,
https://doi.org/10.1007/978-981-19-9319-0_7

bureaucracies that have understood their role as primarily one of *service delivery to* First Nations will have to reorient themselves to become *treaty partners with* First Nations seeking to exercise greater control and autonomy. We cannot yet predict the outcome of these negotiations, nor do we as non-Indigenous writers attempt to articulate First Nations' priorities. However, we argue that a more just policy relationship between treatied First Nations and government is one that, over time, moves towards Indigenous autonomy and self-governance in the place of state authority and control.

The logics of colonialism alongside the efforts of myriad well-meaning individuals seeking to counter them is evident in policy-making. Government departments involve Indigenous people in the work of governance only so far as the arrangement provides the information and reach (connections/networks) that are beneficial to the state. Rarely do these efforts serve Indigenous peoples' broader purposes of restoring and rebuilding their cultures and their ways of life. Despite good intentions by institutions (or individuals within those institutions), what remains is the reality for Indigenous communities of having been dispossessed from their lands, and bunched together with other nations in communities/territories far from their own (living as foreigners) Since colonisation began, First Nations have suffered indignities of being viewed and treated as second class citizens by a people and law that has dominated them through force and then subsequently through coercive tactics aimed at conceding just enough to silence uprising while ultimately retaining full ownership and control in order to have relatively unabated access to Indigenous land. Audra Simpson (2011) articulates the reality faced by Indigenous peoples in settler colonial socieies today: 'Indigenous people did not lay down and die; they persist, and in doing so, they defy all expectations—working resolutely to assert their nationhood and their sovereignty against a settler political formation that would have them disappear or integrate or assimilate' (Simpson A, 2011, p. 212).

Treaty opens a renewed possibility for transformation in these colonial relationships. As treaty processes get underway in Australia we want to outline what we think is needed to be able to change Indigenous-settler relations from relations of domination and control to relationships of equality and genuine partnership. This chapter explores the possibilities and challenges of transforming public policy-making through treaty, arguing that it will take time to re-write the partnership manual and enable genuinely Indigenous-controlled policy to become the new political norm. While the other chapters in this volume have primarily focused on Indigenous innovation and ingenuity in the public policy arena, in this chapter we turn the focus back onto the settler state. In what follows, we argue that the transformations in the policy relationship with First Nations that treaty will opens up will require of the state three distinct but overlapping ways of thinking and working. First, because the negotiation of treaties means the recognition of other sovereign authorities—not authorities to be created (as these have existed for millennia) but authorities that will require space to be exercised alongside the state—state bureaucracies will need to make space for these other sovereigns in rethinking the policy process. Second, state bureaucracies will need to profoundly re-think the ways in which they work with treatied First

Nations, relinquishing control over policy making in favour of more genuine partnerships. Finally, state bureaucracies must commit to the work of respecting First Nations' priorities and to honouring treaty relationships as the guiding principles in policy-making.

## 7.2 Making Space: There Are Other Authorities on This Land

Until quite recently, colonialism was commonly perceived as an historical period that produced the global arrangement of states and distribution of cultures as we know them today. The colonisation of specific Indigenous territories tended to be minimised as a singular, historical event, unpleasant but necessary to produce modern nation-states. In recent decades, however, a growing tide of advocacy and theorising has highlighted a shared experience amongst Indigenous peoples of living under the thumb of colonialism as a *structure*—that is a set of social, political, economic and cultural norms, institutions and systems that are maintained by the governing regime and dominant culture of the state within whose borders an Indigenous nation resides. The fact that colonialism persists in these structures has become particularly evident in settler colonial countries like Canada, Australia, New Zealand, and the United States (see Veracini, 2010; Wolfe, 2006, *inter alia*).

For settler populations, the norms, institutions, and systems of the settler state are broadly accepted as 'natural', necessary and logical mechanisms of governance wielded by sovereign governments. Policy and policy-making are central to these mechanisms. As Elizabeth Strakosch (2015, p. 2) has argued, however, this means that policy is a crucially important site of political encounter and engagement between Indigenous peoples and settler states (Strakosch, 2015, p. 2). Colonial authority over the lives of First Nations is normalised and naturalised through the familiarity of state bureaucracy, while bureaucrats and other non-Indigenous people rarely question the legitimacy of settler state jurisdiction over Indigenous people. Current approaches to policy-making in all policy domains leaves unquestioned the assumption that Indigenous peoples have already been incorporated within the settler regime (Strakosch, 2015, pp. 9, 51).

Yet the mechanisms and the authorities that empower bureaucracies are neither natural nor logical. Treaty-making both reveals and disrupts these assumptions, and potentially paves the way for less colonial policy-making that centres Indigenous authority and control. What is needed to begin such work, however, is a deep and broad deconstruction of sovereignty and governance and the assumptions and principles that underpin these (see Alfred & Corntassel, 2005; Borrows, 2018; Cornell, 2015; Corntassel, 2012; Coulthard, 2014; Kuokkanan, 2019; Maddison, 2019; Mills, 2016; Moreton-Robinson, 2015; Ryser, 2012; Seton, 1999; Simpson, 2014, 2017; Tuhiweh-Smith, 2012; Wilmer, 2002, Wolfe, 2006). As a first step, settler governments and settler colonial societies need to recognise and make space for the other sovereignties that exist in the land such that new partnerships can emerge and enrich the political landscape.

Key to understanding colonialism as a structure is the acknowledgement of other authorities that exist alongside the state's authority. Co-existing sovereignties are a contemporary reality, even as settler states continue to resist this fact. In her 2016 lecture at the Wheeler Centre in Melbourne, Kahnawà:ke Mohawk scholar Audra Simpson made the point that what needs to be contended with is not only the reality of ongoing colonisation, but also the existence of '...other systems of politics; other land tenure systems, normative and philosophical orders that predate this one here' (Simpson, 2016, pp. 7, 45). While the character and characteristics of Indigenous nations change over time, their identities as distinct nations remain strong (Kymlicka, 2000, p. 224).

To understand this dynamic it is helpful to distinguish between 'states' and 'nations'. Nations are ancient, but states are recent political systems placed over geographical areas recognised by international law (Fukurai, 2018). The establishment of the Westphalian international system (with the signing of the Peace of Westphalia treaties in 1648), established states according to a 'nation-state' ideology,[1] rather than drawing boundaries around true existing nations. This forced people to 'conform to the boundaries of conquest' (Griggs & Hocknell, 1995, p. 53) instead of to their national identities.

Nations, thus, are 'housed' within states, which are typically laid over multiple nations (Nietschmann, 1994). While states are administrative systems, a nation is a group of people psychosocially and culturally connected (Fukurai, 2018). They are communities who share a common ancestry, language, cultural homeland, and collectively identify as a people and evolve over time (Bruce, 2015; Nietschmann, 1994; Spicer & Spicer, 1992). Alfred and Corntassel (2005) write that '...the peoplehood concept is a flexible and dynamic alternative to static political and legal definitional approaches to Indigenous identities' (p. 610). Indeed, 'Peoples' is an often-used term in international and political discourse and may be preferred over 'nations', which often incites secessionist anxieties and challenges notions of sovereignty.

This is certainly the case in the United States, Canada, and Australia where the concept of *terra nullius* was used to justify the colonisation of North America (via the Doctrine of Discovery) and the Australian continent, and continues to dictate Indigenous-state relations in these countries.[2] Even though treaties were signed

[1] The false belief that the dominant culture in a geographical area is the only one. In fact, there are only a few examples of states comprising one national identity (Iceland, Vanuatu and Korea are examples). Most states are 'multinational', governing multiple nations within their territories and representing them with one voice in the international community (Bruce, 2015; Nietschmann, 1994; Ryser, 2012). Recent estimates suggest there are approximate 5,000–6,000 nations around the world, making up approximately 18 per cent of the global population (Ryser, 2012, p. 68).

[2] *Terra nullius* is a Latin term meaning 'land belonging to no-one' or unoccupied or empty lands. The legal concept was used to justify colonisation of Australia and North America, via the Doctrine of Discovery. While the 1992 Mabo case overturned *Terra nullius* in Australia, its influence in contemporary Australian law remains. Neither Canada nor the United States have revoked the Doctrine of Discovery.

between early settlers and Indigenous nations in Canada, the US, and New Zealand (Australia is the only British-settler majority country to not have a history of treaty-making), these have subsequently been interpreted in such a way to affirm colonial control rather than as commitments to relationships between multiple sovereignties in a shared territory. Since the early colonial period, the colonial projects in Australia, New Zealand, and North America have sought to eliminate Indigenous nations in an effort to secure access to land and resources (Wolfe, 2006) through a singular sovereignty. However, as has been thoroughly covered in the literature, this work has been unsuccessful (Simpson A, 2011, p. 212). First Nations still exist, along with their sovereignties and the rights and responsibilities this authority holds.

Contemporary treaty-making rests on the recognition of these other sovereign entities with whom the settler state might negotiate agreements. The concept of 'recognition' however is also problematic. Critical Indigenous scholars including Glen Coulthard (2014), Leanne Betasamosake Simpson (2017) and Audra Simpson (2014) raise the question: who is recognising whom? And by what authority do they recognise another? Audra Simpson (2014, p. 662) observes that 'political recognition is, in its simplest terms, to be seen by another *as one wants to be seen.*' To recognise First Nations as sovereign, self-determining bodies (in accordance with international law and in response to decades of advocacy), however, poses a dilemma for settler colonial societies, which cannot proffer this recognition and continue to control all aspects of Indigenous life. At the same time, Indigenous identities and the rights and responsibilities these identities hold, cannot be expressed or understand through colonial language or within the context of colonial systems and worldviews (Simpson, 2004). This dilemma strikes at the heart of the 'politics of recognition', which Coulthard (2014, p. 3) condemns as tactics by the state to maintain control: '…instead of ushering in an era of peaceful coexistence grounded on the ideal of reciprocity or mutual recognition, the politics of recognition in its contemporary liberal form promises to reproduce the very configuration of colonialist, racist, patriarchal state power that Indigenous peoples' demands for recognition have historically sought to transcend.' Contemporary efforts to recognise and 'accommodate' First Nations' political authority still assume the domestication of sovereign nations (Alfred & Corntassel, 2005, p. 604).

True recognition, then, is not about one sovereign entity defining the other. Rather, it is the acknowledgement that another sovereign authority exists, a fact that requires substantive change in the relationship in order to facilitate peaceful coexistence. While traditional definitions of sovereignty as monopolistic and static (Besson, 2011) would not allow this kind of recognition, more recent understandings of sovereignty suggest that it is more fluid and divisible and applies both internally, within territorial boundaries, and externally, between states (Brennan et al., 2004, p. 310). As the former Australian Aboriginal and Torres Strait Islander Social Justice Commissioner William Jonas (2002) explains, sovereignty 'is constantly being re-aligned and redistributed amongst myriad levels and players'. Broadly, sovereignty is increasingly understood to be enshrined in people, rather than institutions (Brennan et al., 2004, pp. 310–311; Hobbs & Williams, 2018, p. 4), meaning that where peoples exist, so does their sovereignty. Settler colonial societies are contending with this reality in various ways, including through reforms in agreement-making, land-titling, power-sharing and co-management arrangements, consultations, and policy co-design. Treaty offers the possibility of more substantive recognition of First Nations sovereignties, with significant implications for public policy.

When Europeans arrived in Australasia and North America, the lands were not empty. Nor were the British or subsequent settler populations successful in eliminating original nations. Narratives of space (such as *terra nullius*) and time (the erasure of Indigenous histories) were used by states as colonial tools, squarely placing states as centres of power and identity in the story of human history (Seton, 1999, p. 5). These narratives shaped settler culture and established the foundations of contemporary governance and public life in these countries. Indigenous social, political, and economic systems and ontologies have been severely harmed by colonialism, but they continue, and in some cases are gaining in strength as First Peoples take steps to rebuild their nations. As such, treaties will not create new authorities, but they must create space for these existing authorities to be exercised alongside that of the state. Making space, in this sense, involves a shift from a normative neoliberal recognition framework to an engagement with the co-existing sovereignties that underpin Indigenous-settler relations. This will require both the state's financial and long-term investment in Indigenous nation-building projects (as a reparative responsibility), and a fundamental reimagining of the terms of its own sovereignty from an outdated, Westphalian monolith, to something more complex and indeed more bounded. This will not be a sudden change. In the hypercolonised spaces of settler states—in both the geographical and political sense—time is needed for Indigenous peoples to regenerate their nationhood while states need to unlearn and lay down their colonial tools.

## 7.3 Partnering: Moving from Control to Cooperation Through Cooperative Governance

As treaty processes in Australia begin to open new possibilities for more balanced and just relationships between settler governments and First Nations, the 'partnership manual' that governs the relationship will need to be rewritten. While some efforts have been made in this regard through the introduction of participatory processes and increased consultations, these practices remain firmly situated within the structures and control of state bureaucracy (Curran, 2019, p. 2; Von der Porten et al., 2015, pp. 137–138). First Nations are invited to participate in these processes but have no influence over the rules of the game. Early treaty-making practices in other settler colonies offer some insight into what more balanced relationships between the state and Indigenous nations can look like. However, the subsequent interpretation of these treaties by settler governments has obscured these early agreements under a colonial patina.

One emerging area of promise where a new policy relationship dynamic can be explored is cooperative governance (Curran, 2019; Von der Porten et al., 2019a; Wyatt & Nelson, 2013; Zurba, 2014). While cooperative governance arrangements are also largely established and managed within the firm grip of the state, Indigenous nations are finding ways to assert their authority and push the state into more equitable relationships reflective of early treaties. What is clear, is that alternatives to the status quo partnership model will not be unilaterally imagined by the state. It is only through the practice of partnering—including experimentation, empowering Indigenous sovereignty, and unpacking and dismantling colonial norms—that a just future can be realised.

In Australia, the concept of Indigenous political authority and right to self determination has largely been viewed as a threat to state sovereignty. To date, and not withstanding the emerging treaty processes noted above, there is no formal recognition of Indigenous sovereignty or rights nor agreement-making processes to define or distribute authority. The 1992 Mabo v Queensland (No. 2) decision affirmed the existence of native title, launching a national native title process through which Indigenous groups can gain some rights to occupy and use their traditional lands. The case also found that sovereignty cannot be considered by domestic courts of the nation whose sovereignty is in question (Brennan et al., 2004, pp. 318–319). Three additional High Court decisions rendered Indigenous sovereignty non-justiciable: Coe v Commonwealth (No. 2) [1993]; Commonwealth v Yarmirr [2002]; Members of the Yorta Yorta Aboriginal Community v Victoria [2002]. However, all three positions on Indigenous sovereignty are contestable given their acknowledgement that sovereignty is fluid and must be considered in context, that Indigenous legal systems are still exercised and that Indigenous peoples were sovereign before European arrival (Brennan et al., 2004, p. 328).

Rather than being seen as a right and as a framework for advancing First Nations' authority and autonomy, in Australia self-determination is treated as a principle to inform policy and program design and implementation by the state. As in the other settler states, First Nations and Australian governments build and maintain partnerships focused on land and natural resources co-management principles. However, these partnership exist only in accordance with the structures, rules, and timeframes set out by the state. While there have been legal reforms that have allowed for greater political participation by Aboriginal and Torres Strait Islander peoples and some recognition of their unique rights, these changes have not amounted to a 'meaningful amendment to the framework of governance' (Hobbs, 2018, p. 315). Harry Hobbs (2018) argues that the Australian federation has largely failed to recognise and provide space for Indigenous self-determination as it has been expressed by Aboriginal and Torres Strait Islander peoples. This is in part because of the system's 'unitary demos'. In other words, it is not structured to respond to diversity (Hobbs, 2018, p. 324), particularly with regard to co-existing sovereignties.

Within this context of limited tolerance for notions of Indigenous sovereignty, Indigenous nations have few options for the exercise of their right to self-determination. So called self-governance arrangements are crafted and carefully managed by governments to ensure they continue to serve the objectives of the state. To do so, settler state governments draw from a well-established toolbox of tactics for engaging with Indigenous peoples that give the appearance of providing for autonomy and partnership, while in fact maintaining the status quo of control over Indigenous peoples. Some of the common tactics that we have observed in these relations include:

- *Task flooding—the Indigenous group is buried in tasks and deliverables, but has no actual authority.* This involves giving significant responsibilities with no authority over the structure (the 'what' or 'how') in which those responsibilities are occurring. The First Nation's capacity is stretched with implementing the state's program, including meeting reporting requirements, undertaking training, participating in consultations or on advisory boards and various other administrative activities that it has not time or resources to commit to developing and advocating for its own structures, processes and priorities. This is observable in co-management schemes in which Indigenous groups partner with a government department or agency around a particular objective, generally land and resource management.
- *Sidelining—keeping Indigenous people out of the rooms in which decisions are made.* This tactic includes neglecting to invite Indigenous people to meetings, failing to keep them in the loop or not sharing information about policy or programs that affect them and their communities. In some cases, a select few representatives may be invited to participate but it is rare that space and time are provided for engaging with a diversity of views amongst Indigenous communities and individuals. Where these views are voiced and documented, they are interpreted and applied by the state bureaucracy. This is a standard approach to community consultations by settler state governments.
- *Undermining (the gap between responsibility and resources)—verbally giving responsibility to Indigenous groups but then not supporting them, changing the rules, or overriding or blocking their decisions.* In an effort to support self-determination, governments will appoint an Indigenous group or individual as the lead on an issue, but then not support their decisions or provide them with the resources they need to follow through. Other undermining tactics include changing the rules or policies (requirements, funding, timelines, procedures etc.) concerning the issue or overriding or blocking decisions or actions taken by the Indigenous group/individual leading the process. An infamous example of this is the rejection of the Uluru Statement from the Heart by the Turnbull government in 2017.
- *Failure crafting—setting circumstances under which it is near impossible for the Indigenous nation to succeed.* Related to 'undermining', 'failure crafting' occurs when an Indigenous group or individual is responsible for achieving an enormous, complex task with inadequate resources or support or within an unrealistic timeframe. This tactic is often evident in community governance schemes where some form of devolution of authority and resources is involved. Smith (2002, p. 5) refer to the 'dump and run' approach Australian governments have historically taken

> with self determination initiatives, whereby following the devolution of power and resources 'many government departments and non-government agencies have "vacated the field", withdrawing staff and practical support'.

Both the structures and practices of settler state governments are oriented toward maintaining state dominance and control over Indigenous peoples. While Indigenous sovereignty and rights are gifted significant rhetoric, there is very little action that genuinely contends with the notion of other sovereign authorities in a shared space.

It is hoped that treaty will be an opportunity to radically revise both the infrastructure and the rules that house and govern Indigenous-state relations in Australia. Ideally, treaty will change the political and policy relationship such that one partner no longer holds all the power and resources with a modus operandi governed by principles of domination and control. Just what these changes will look like in practice remains an open question; one that is being wrestled with by scholars, advocates, and practitioners across sectors and disciplines around the globe. So long as states remain the sole internationally recognised units of sovereign power, even treatied First Nations will need to prioritise their nationhood rebuilding, nurture their sources of strength, and creatively draw on myriad sources of support to survive their encounters with the state. What is being called for is respectful relationships with governments that are not characterised by domination. This reorientation from the troubled coexistence between First Nations and the state in Australia, to a relationship in which the parties share equal influence and draw equal benefit requires 'rebuilding the intellectual resources and political space to have more symmetrical, reciprocal and respectful conversation within and between Indigenous and state legal traditions' (Napoleon & Friedland, 2014, p. 15). This process of 'rewriting the partnership manual' is multifaceted and will look different for each government and Indigenous nation. First Nations are seeking respect for their autonomy (as the nation defines it) and sovereign authority, in the context of Indigenous-state relationships that are guided by the principle of reciprocity.

Central to any future transformation in the policy relationship between First Nations and settler states will be the need for states to yield to Indigenous processes and priorities rather than merely inviting Indigenous people to participate in state-controlled work. As noted above, settler governments have myriad tactics of control that they employ in their interactions with Indigenous peoples. In recent decades some governments have adopted various participatory and 'co-design' approaches ostensibly intended to increase Indigenous participation in government processes. However, these co-design processes still exist wholly within state machinery and are designed and controlled by government bureaucracy. Scope, timeframes, objectives and funding parameters are often decided behind closed doors, and negotiations with Indigenous partners on these elements are often limited by parameters set by the state. Indeed, while these processes may appear to be collaborative and equal, the contexts in which they occur are biased structurally toward state interests. It is prudent to note that many individual bureaucrats and government representatives work tirelessly to engage with Indigenous peoples on balanced footing and in good faith. However, no amount of good intention in these processes can pacify the colonial principles that are deeply embedded within the structures of settler governance.

With increasing national and international advocacy for self-determination and self-governance, many governments have and continue to explore political arrangements with Indigenous nations that would accommodate some degree of autonomous governance. These arrangements are generally in the form of cooperative governance arrangements, practices, or mechanisms (also referred to as 'multilevel governance') which can include horizontal and/or vertical power-sharing, shared decision-making, spatial arrangements, pluralistic law-making, negotiations protocols and co-management (Daniell & Kay, 2017; Hawkes, 2001; Ivison et al., 2000; Papillon, 2011; Ryser, 2012; Smith, 2002). Cooperative governance arrangements provide for two or more governing authorities to exercise their powers in concert with one another. A significant complexity of cooperative models is the continuous coordination and negotiation of resource and power-sharing between the two jurisdictional authorities (Daniell & Kay, 2017). Cooperative arrangements are the 'venues through which [Indigenous groups] interact with federal and state/provincial authorities' (Papillon, 2011, p. 294). While these venues and practices vary from one political context to the next, the literature broadly describes cooperative or multilevel governance as 'negotiation systems'—a patchwork of informal and formal structures and practices of consensus building and accountability (Daniell & Kay, 2017).

By design, cooperative governance arrangements provide First Nations and states with an avenue to negotiate and practice their political relationship. While states seek to employ the same control and extraction tactics as in other areas of Indigenous-state relations, First Nations (particularly in Canada) are using cooperative arrangements to model and work toward a more just political relationship with the state. For instance, a study of water governance in British Columbia has found that First Nations are establishing parallel governing systems based on their traditional knowledge and systems as a means of countering embedded colonial governance practices and shifting the balance of power between themselves and the provincial and federal authorities (Curran, 2019, p. 1). Looking specifically at ground water licencing, Curran (2019) observes that 'the state depoliticizes decisions about water by directing it into administrative processes like environmental assessment' (p. 2). First Nations are invited to participate in governance through administrative mechanisms that provide no opportunities for dialogue about the governance structure itself. Any concerns that Indigenous representatives have with the functioning of the project are minimised to administrative issues dealt with through an assessment process or consultations. In response, Indigenous leaders '…are repoliticising water governance by creating evaluative processes that reflect their own legal traditions and standards', justifying their side-stepping of state governance with their own legal authority and the UNDRIP:

> They are redefining political and governance processes by declaring their own Indigenous laws that mandate ecological and procedural outcomes. They are inserting new governance processes into state administrative activities beyond the narrow mandate of 'consultation and accommodation' that serve to repoliticise water governance using the language of consent. (Curran, 2019, p. 7)

Another Canadian study found that First Nations are innovating new governance approaches in natural resource management, introducing models of cooperative governance with the state that affirm Indigenous authority and resemble more equitable governing partnerships (Von der Porten et al., 2019b, pp. 528–530). Broadly, First Nations are increasingly declaring their legal authority and asserting their rights over their water, lands, and resources in accordance with the UNDRIP. They are setting 'expectations for behaviour and responsive action within their traditional territories' (Curran, 2019, pp. 7–8). In a study of Haida, Nuu-chah-nulth and Heiltsuk nations' engagement with Canadian Department of Fisheries and Oceans (DFO), Von der Porten et al. (2019a) identify various strategies used by the First Nations to assert their right and authority including issuing Indigenous declarations and bans, negotiating directly with industry, and writing and issuing management plans (p. 66).

The manoeuvres employed by Indigenous nations show the way for a different kind of relationship with settler states. As Barkandji scholar Zena Cumpston argues, 'Our community leaders must not only be given a seat at the table, they should set the menu too' (Cumpston, 2020). This means the terms and structures of the relationship are determined together—the *why* (purpose) of the interaction is clear, and the *how* (structures, processes and timeframes) of the relationship is jointly and fairly agreed. However, because of the impacts of colonisation, not all Indigenous nations are currently in a position to engage in a political dialogue with the state. This is either due to other pressing community priorities, limits on capacity or out of self-preservation, recognising that interactions with the state tend to be costly (Simpson, 2016, p. 24). Many Indigenous nations are turning inward to necessarily focus on their own nationhood building priorities. This will not magically change with the signing of treaties, and honouring treaty relationships with First Nations must include time and space for this work to unfold.

## 7.4 Honouring: Upholding Treaty Relationships in Policy-Making

As in any relationship, the health of the partnership hinges on the health of the parties. To date, the relationship between settler states and First Nations residing within their borders has been characterised by abusive tactics of domination and control. For generations, Indigenous nations have fought to survive under a torrent of assault on their identity, both as individuals and particularly as nations. If a different kind of relationship is to be realised, the first step must be to lift the pressures of domination and distraction burdening Indigenous nations and hold space for them to revive their languages, cultural practices, connections to their lands, governance, kinship and intelligence systems, before, during and after the negotiation of treaties. From this place of renewed strength, it may be possible for First Nations and settler governments to together create alternatives to the colonial relationship.

The work that governments must necessarily undertake is to acknowledge and dismantle the colonial principles that underpin state systems and the practices that serve to maintain the abusive power dynamic. This is the work of centering Indigenous priorities and honouring treaty relationships, where they exist, as the guiding principles in policy-making. One significant aspect of this is the need for states to 'hold space' for First Nations engaged in the necessary work of reviving their own ontological, governing systems and social structures, rather than demanding of Indigenous peoples that they work according to timelines set by the settler state and its electoral and budget cycles.

In response to the ongoing assault on Indigenous identities, connections and ways of life, there is a growing call among First Nations to focus their energies on rebuilding Indigenous cultures, institutions, systems and knowledge traditions (Elliott, 2018; Simpson L, 2011). This work is referred to as both Indigenous 'refusal' and 'resurgence' and involves 'decentering' the state in the lives and work of First Nations and instead returning focus to relationships with human and non-human communities and land and water-based governance through daily acts of resurgence (actions that are the 'embodiment of self-determination'). Audra Simpson (2014, p. 11) describes refusal as 'a political and ethnic stance that stands in stark contrast to the desire to have one's distinctiveness as a culture, as a people, recognised.' Rather than seeking recognition, in practicing refusal, Indigenous people turn away from the settler state to center their own values, principles, social structures and governing principles.

Relatedly, resurgence also focuses on the revival of Indigenous ontologies independent of the state. Resurgence operates at several levels, and is first a personal process for each Indigenous person, of decolonising one's own thinking and behaviour (Coulthard, 2014; Simpson L, 2011). First, live according to the ethics of your nation in your personal life, evaluating attitudes, actions and choices according to that intelligence. Second, from that place of personal transformation, build a movement to develop an alternative way of relating to the settler system (Simpson, 2016). Jeff Corntassel (2012) describes this process as acts of 'everyday resurgence': 'How one engages in daily processes of truth-telling and resistance to colonial encroachments is just as important as the overall outcome of these struggles to reclaim, restore, and regenerate homeland relationships' (p. 89).

The acts of resurgence that follow individual decolonisation are a collective 'resurgence mobilisation'; the process of looking inward to Indigenous traditions and breathing life into those again. For Leanne Simpson (2011), this starts with Anishinaabeg creation stories, which are the ontology of Anishinaabeg intellectual tradition or, quoting Coulthard, Anishinaabeg 'grounded normativity' (Simpson L, 2011). Priority actions of Indigenous nationhood resurgence are connecting with land, practicing culture, renewing relationships with community, engaging in Indigenous diplomacy, connecting with other resistance communities, re-establishing governance and governance systems (Elliott, 2018). This is a process of critical traditional systems resurrection. Indigenous knowledge is accessed, shared, and preserved through ceremony, dance, song, relationships with each other and the land, and storytelling (Simpson L, 2011). Alfred and Corntassel (2005) offer a pathway to decolonisation and resurgence for Indigenous peoples: reconnect with land, learn and use Indigenous language, investigate and challenge complacency, 'decolonize your diet' and build capacity for self-sufficiency and reinstitute the mentor-learner relationship teaching methodology.

Practices of refusal and resurgence connect with and support the work of Indigenous nation (re)building. Cornell (2015) observes that there are typically three stages that Indigenous nations move through in order to become self-governing and ready to develop cooperative arrangements with non-Indigenous governments: First, they must identify as a nation and the nation's members need to view themselves as a cohesive unit. The representative functions of the nation need to be structurally sound and have the capacity to undertake governance functions and be accepted as legitimate by the nation's members. Second, they must organise as a nation and establish their own political and institutional structures and processes. Third, they must act as a nation and engage with the state in an intergovernmental relationship by articulating clearly what it wants and needs, managing the implementation of policies or strategies and leading the monitoring and review processes (Cornell, 2015).

The role of the settler state in this work of resurgence and rebuilding is to hold space for First Nations to evolve in their own way and time. This means, not interfering and standing ready to support *if and how* the Nation asks. The work of Indigenous nation (re)building is the right and responsibility of the nation itself. Many First Nations have been undertaking this work for years, away from the public eye (see Cornell, 2015; Rigney et al., 2021, *inter alia*). However, for nations wanting to benefit where they can from state offerings, engagement with state bureaucracy is inevitable, tying up leadership and governance resources that could otherwise be directed at nation building work. For governments and organisations, support for Indigenous nations engaged in rebuilding processes may be things like providing funding with no KPIs or reporting requirements, running programs for the community so they can focus on their own priorities, and amending plans and legislation to enable Indigenous people to access land and waterways for ceremonial and related purposes. The work of rebuilding nations and recentering nationhood is long and necessary for generating alternatives to the colonial relationship with the state. Leanne Simpson (2017) argues that the revival or resurgence of Indigenous nationhood is not about dismantling or holding the state accountable for its assault against Indigenous peoples. Rather, it is about the regeneration of Indigenous ways of being, and in that, the creation of alternatives to the colonial relationship between Indigenous nations and settler states. As Simpson contends, 'I'm not particularly interested in holding states accountable because the structure, history and nature of states is exploitative by nature. I'm interested in alternatives. I'm interested in building new worlds' (Simpson, 2017, p. 31).

Treaty relationships are a mechanism through which alternatives to the colonial relationship might be realised. However, this can only happen if both Indigenous nations and governments are prepared to build new relationships on different terms. The work that Indigenous nations face will be focused on recentering, strengthening, and connecting. The state, on the other hand, must contend with its coloniality and consider what needs to be dismantled so that a new Indigenous-settler political landscape can emerge. In reorienting to treaty relationships with First Nations in this way, settler governments will need to 'unlearn' ways of relating to Indigenous peoples that are centred on individualism, extraction and domination. Key values like individualism, land as property, and even liberal democracy itself need to be interrogated to reveal the ways in which they maintain the colonial relationship between states and Indigenous peoples. These are elaborated on below.

The liberal democratic value of individualism pervades governance in settler states. Broadly speaking, states prefer to develop policy focused on Indigenous peoples as collections of individuals, focusing on improving their socio-economic outcomes while ignoring their political identity as nations. Indigenous nations on the other hand, do not ignore welfare issues but rather view them in the context of their collective well-being. Yet the state experiences the resurgence of Indigenous collectives in the form of First Nations as threatening and difficult to accommodate. By imposing policy 'solutions' onto Indigenous peoples based solely on the principle of individuality, governments deny collective Indigenous identities and aspirations (Cornell, 2015, p. 5). In this context, Indigenous nationhood is essentialised as a means of countering the threat of an empowered people. Indigenous culture and the right to self-determination may be given rhetorical attention, but do not meaningfully inform how the state makes decisions and policies.

A key stage on which the Indigenous-state relationship unfolds concerns the governance of land. Relationships between the state and Indigenous nations hinge on the question of land. As discussed earlier, land is central to Indigenous sovereignty, governance, spirituality, and intellectual traditions. It is the source and living relationship that sustains and generates Indigenous lifeways. In entering into treaty relationships with Indigenous peoples the state must be prepared to undergo a fundamental shift in its own orientation to the lands within their boundaries. Through colonisation, 'Land is remade into property and human relationships to land are restricted to the relationship of owner to his property' (Tuck & Yang, 2012, p. 5). Treaties that perpetuate the orientation toward land as property are simply further tools of colonisation. So long as the settler system persists with its property/commodity posture toward land and the natural world, it will prioritise destructive practices over relational ones, and Indigenous peoples will continue to be dispossessed of that connection. In order for treaties to represent renewed relationships with Indigenous peoples and to allow for the genuine exercise of the right to self-determination, they must necessarily empower the restoration of human-land relationships for Indigenous peoples. This includes provision for the exercise of Indigenous sovereignty over lands and therefore enabling Indigenous peoples' reciprocal relationship orientation to prevail over that territory.

Indigenous self-determination and self-governance rights can be interpretated in such a way as to either stifle or enable the exercise of Indigenous legal and governing systems (Starblanket, 2019, p. 13). All four settler states reference these rights in policy and legislation, and yet none has succeeded in their alleged commitment to recognising, protecting, and empowering Indigenous ways of governing in action. Instead, in their self-governance accommodation schemes, settler governments typically impose western liberal democratic ideas of 'good governance' onto Indigenous nations, meaning Indigenous governance bodies are often required to mirror those of the state. Governments view this as essential for the harmonisation of political systems. However, this restriction on Indigenous governance continues to hamper the evolution of Indigenous governing traditions and the development of culturally relevant institutions and processes, which evidence shows are critical to the social and economic well-being of First Nations (Cornell, 2002), as well as decolonised relationships between First Nations and states (Cornell, 2015).

Developing new relationships based on reciprocity will require a deep investigation of the principles and practices that guide state policy and decision-making. Treaties are not the end goal in this process. They are neither a conclusion to colonisation nor a definitive statement of the relationship between governments and First Nations. However, as a commitment to learning and nurturing a relationship of reciprocity, treaty relationships offer an alternative to colonial principles that can guide policy and decision-making toward a more just future.

## 7.5 Conclusion

This chapter has outlined some of the opportunities and challenges that treaty creates for the move towards Indigenous control of policy that affects the lives of First Nations people. We have argued that for treaty to provide a pathway to Indigenous-controlled policy the settler state will be required to make space for Indigenous sovereignties; conceive of new ways of partnering with First Nations; and must honour the new relationships that will be created through treaty, allowing time and space for First Nations to change and develop their own governance capacities over time. All of this will take time, and the state and its agencies will have to learn a vastly different way of working—from the political to the bureaucratic, the individuals and institutions that make up the state will need to adjust their expectations, lay down their current toolsets and pick up and learn how to use new ones. If the goal is a balanced, just relationship between First Nations and the state, the burden of responsibility to evolve does not lie on Indigenous peoples. The colonial state will not be dismantled through a single piece of legislation or through one dramatic decision or declaration. Changing the way the state works will be incremental and will require education, reform, and a persistent chipping away of the dominating tendences that are part of the settler state's DNA.

Decolonisation and resurgence writers are challenging settler societies and state governments to think outside of the box. Notions of land-based alternatives to capitalism, political relational autonomy, power redistribution, and multinationalism all offer alternatives to neoliberalism. These may seem radical, far-fetched ideas, but as has been discussed, they are not new. In fact, these notions are no more radical than those faced by previous generations determined to address relations of inequality and oppression—the abolition of slavery, work toward gender equality, and rights for children, for example, were all revolutionary ideas that seemed inconceivable by populations of the time.

But as this work begins—and indeed it already has begun—new norms of governance will develop. The social movements and political reforms of the past have taught us that we can learn and adjust to new social and political norms. Before arriving at those norms, the prospect of change seems overwhelmingly daunting and unattainable. But we are on this new path. Treaty is opening space for new/old possibilities, new relationships, and a radically different shared future in which First Nations autonomy and control over policy is an expectation rather than an exception.

## References


Alfred, T., & Corntassel, J. (2005). Politics of Identity—IX: Being indigenous: Resurgences against contemporary colonisation. *Governance and Opposition, 40*(4), 597–614.

Besson, S. (2011). Sovereignty. In *Oxford public international law*. Retrieved January 17, 2022, from https://opil.ouplaw.com/view/10.1093/law:epil/9780199231690/law-9780199231690-e1472

Borrows, J. (2018). Earth-bound: Indigenous resurgence and environmental reconciliation. In M. Asch, J. Borrows, & J. Tully (Eds.), *Resurgence and reconciliation: Indigenous-settler relations and earth teachings*. University of Toronto Press.

Brennan, S., Gunn, B., & Williams, G. (2004). 'Sovereignty' and its relevance to treaty-making between indigenous peoples and Australian governments [online]. *Sydney Law Review, 26*(3), 307–352.

Bruce, H. (2015). Fourth world theory and state collapse. *Fourth World Journal, 13*(2), 61–66.

Cornell, S. (2002). Nation-building and the treaty-process. *Indigenous Law Bulletin, 5*(17), 1–6.

Cornell, S. (2015). Processes of native nationhood: The Indigenous politics of self-government. *International Indigenous Policy Journal, 6*(4), 1–27.

Corntassel, J. (2012). Re-envisioning resurgence: Indigenous pathways to decolonization and sustainable self determination. *Decolonization: Indigeneity, Education and Society, 1*(1), 86–101.

Coulthard, G. (2014). *Red skin, white masks: Rejecting the colonial politics of recognition*. University of Minnesota Press.

Cumpston, Z. (2020). To address the ecological crisis Aboriginal peoples must be restored as custodians of country. *The Conversation*. https://theconversation.com/to-address-the-ecological-crisis-aboriginal-peoples-must-be-restored-as-custodians-of-country-108594

Curran, D. (2019). Indigenous processes of consent: Repoliticising water governance through legal pluralism. *Water, 2019*(11), 1–16.

Daniell, K. A., & Kay, A. (2017). Multilevel governance: An introduction. In K. A. Daniell & A. Kay (Eds.), *Multi-level governance* (pp. 3–32). ANU Press.

Elliott, M. (2018). Indigenous resurgence: The drive for renewed engagement and reciprocity in the turn away from the state. *Canadian Journal of Political Science/revue Canadienne De Science Politique, 51*(1), 61–81. https://doi.org/10.1017/S0008423917001032

Fukurai, H. (2018). Fourth world approaches to international law (FWAIL) and Asia's indigenous struggles and quests for recognition under international law. *Asian Journal of Law and Society, 5*(1), 221–231.

Griggs, R., & Hocknell, P. (1995). The geography and geopolitics of Europe's fourth world. *Boundary and Security Bulletin, 3*(4), 59–67.

Hawkes, J. (2001). *The fourth pillar of sustainability: Culture's essential role in public planning*. Common Ground.

Hobbs, H. (2018). Aboriginal and Torres Strait Islander peoples and multinational federalism in Australia. *Griffith Law Review, 27*(3), 307–336. https://doi.org/10.1080/10383441.2018.1557587

Hobbs, H., & Williams, G. (2018). The Noongar Treaty Settlement: Australia's First Treaty? *Sydney Law Review, 40*, 1–42.

Ivison, D., Patton, P., & Sanders, W. (Eds.). (2000). *Political theory and the rights of indigenous peoples*. Cambridge University Press.

Jonas, W. Dr. Aboriginal and Torres Strait Islander Social Justice Commissioner. (2002). *Recognising aboriginal sovereignty—Implications for the treaty process* [Speech transcript]. Australian Human Rights and Equal Opportunity Commission. https://humanrights.gov.au/about/news/speeches/recognising-aboriginal-sovereignty-implications-treaty-process-2002

Kuokkanan, R. (2019). 'Self-determination: Foundational value'. In *Restructuring relations: Indigenous self-determination, governance, and gender*. Oxford University Press.

Kymlicka, W. (2000). In D. Ivison, P. Patton, & W. Sanders (Eds.), *Political theory and the rights of indigenous peoples*. Cambridge University Press.

Maddison, S. (2019) *The colonial fantasy: Why white Australia can't solve black problems*. Allen & Unwin. https://discovery.ebsco.com/linkprocessor/plink?id=d3eb57f3-941a-351c-a042-683120595603. Accessed 16 Mar 2022.

Mills, A. (2016). The lifeworlds of law: On revitalizing indigenous legal orders today. *McGill Law Journal/revue De Droit De McGill, 61*(4), 847–884.

Moreton-Robinson, A. (2015). *The white possessive: Property, power, and indigenous sovereignty*. University of Minnesota Press. https://discovery.ebsco.com/linkprocessor/plink?id=54316637-56ef-398d-8c65-e7caf22aa5e5. Accessed 16 Mar 2022.

Napoleon, V., & Friedland, H. (2014). Indigenous legal traditions: Roots to renaissance. *The Oxford handbook of criminal law* (pp. 225–247). Oxford University Press.

Nietschmann, B. (1994). The fourth world: Nations versus states. In G. Demko & W. B. Wood (Eds.), *Reordering the world* (pp. 225–237). Westview Press.

Papillon, M. (2011). Adapting federalism: Indigenous multilevel governance in Canada and the United States. *The Journal of Federalism, 42*(2), 289–312. https://doi.org/10.1093/publius/pjr032

Rigney, D., Bell, D., & Vivian, A. (2021). Talking treaty: A conversation on how indigenous nations can become treaty ready. In H. Hobbs, W. Alison, & L. Coombes (Eds.), *Treaty-making: Two hundred and fifty years later*. The Federation Press.

Ryser, R. (2012). *Indigenous nations and modern states: The political emergence of nations challenging state power*. Routledge.

Seton, K. (1999). Fourth world nations in the era of globalisation: An introduction to contemporary theorizing posed by Indigenous nations. https://nointervention.com/archive/pubs/CWIS/fworld.html

Simpson, A. (2011). Settlement's secret. *Cultural Anthropology, 26*(2), 205–217. https://doi.org/10.1111/j.1548-1360.2011.01095

Simpson, A. (2014). *Mohawk interruptus: Political life across the borders of settler states*. Duke University Press.

Simpson, L. (2004). Anticolonial strategies for the recovery and maintenance of indigenous knowledge. *American Indian Quarterly, 28*(3/4), 373–384.

Simpson, L. (2011). *Dancing on our turtle's back: Stories of Nishnaabeg re-creation, resurgence and a new emergence*. Arbeiter Ring Publishing.

Simpson, A. (2016). *Public lecture*. The Wheeler Centre, 22 February.

Simpson, L. (2017). *As we have always done*. University of Minnesota Press.

Smith, D. (2002). *Jurisdictional devolution: Towards an effective model for indigenous community self-determination* (Discussion paper No. 233/2002). Centre for Aboriginal Economic Policy Research, ANU.

Smith, L. T. (2012). *Decolonizing methodologies: Research and Indigenous peoples*. Zed Books. https://discovery.ebsco.com/linkprocessor/plink?id=d29f0ca1-e59d-3fbc-999f-58fe8361cb2f. Accessed 16 Mar 2022.

Spicer, E., & Spicer, R. (1992). The nations of a state. *Boundary 2, 19*(3), 26–48.

Starblanket, G. (2019). Constitutionalizing (in)justice: Treaty interpretation and the containment of indigenous governance. *Constitutional Forum Constitutionnel, 28*(2), 13–24.

Strakosch, E. (2015). *Neoliberal indigenous policy: Settler colonialism and the 'post-welfare' state*. Palgrave Macmillan. https://discovery.ebsco.com/linkprocessor/plink?id=22f083fe-f4fa-3340-9c29-a7d1cd80256b. Accessed 16 Mar 2022.

Tuck, E., & Yang, W. (2012). Decolonization is not a metaphor. Decolonization: Indigeneity. *Education & Society*, *1*(1), 1–40.

Veracini, L. (2010). *Settler colonialism: A theoretical overview*. Palgrave

Von der Porten, S., de Loë, R., & Plummer, R. (2015). Collaborative environmental governance and indigenous peoples: Recommendations for practice. *Environmental Practice 17*(2), 1–11.

Von der Porten, S., Corntassel, J., & Mucina, D. (2019a). Indigenous nationhood and herring governance: Strategies for the reassertion of indigenous authority and inter-indigenous solidarity regarding marine resources. *Alternation, 15*(1), 62–74.

Von der Porten, S., Ota, Y., Cisneros-Montemayor, A. & Pictou, S. (2019b). The role of indigenous resurgence in marine conservation. *Coastal Management 47*(6), 527–547.

Wilmer, S. E. (2002). *Theatre, society, and the nation: Staging American identities*. Cambridge University Press. https://discovery.ebsco.com/linkprocessor/plink?id=59b2b451-57f0-38b2-adac-e33f2742275e. Accessed 16 Mar 2022.

Wolfe, P. (2006). Settler colonialism and the elimination of the native. *Journal of Genocide Studies, 8*(4), 387–409.

Wyatt, S., & Nelson, H. (2013). Aboriginal engagement in Canada's forest sector: The benefits and challenges of multilevel and multi-party governance. In M. Papillon & A. Juneau (Eds.), *Canada: The State of the Federation 2013: Aboriginal multilevel governance* (pp. 119–142). McGill-Queen's University Press.

Zurba, M. (2014). Leveling the playing field: Fostering collaborative governance towards on-going reconciliation. *Environmental Policy and Governance, 24*, 134–146. https://doi.org/10.1002/eet.1631

# Chapter 8
# Yes, The Time Is Now: Indigenous Nation Policy Making for Self-determined Futures

**Miriam Jorgensen, Alison Vivian, Anthea Compton, Donna Murray, Debra Evans, and Janine Gertz**

**Abstract** There is a proliferation of ideas about the meaning and content of the term 'Indigenous nation building' (INB). Returning to the findings of the Harvard Project on American Indian Economic Development, INB refers to the exercising of decision-making authority and strengthening of a nation's capacity for legitimate governing, which is associated with achieving a nation's self-determined goals. In Australia, a number of Aboriginal and Torres Strait Islander nations are doing exactly this. The Gunditjmara, Gugu Badhun, and Wiradjuri nations are working to rebuild their governing foundations, strengthen community governance and, in so doing, advance nation goals. These nations have been able to do so not because of settler colonial policy support, but because of their own Indigenous public policy. The Gunditjmara, Gugu Badhun and Wiradjuri nations have worked to seize available opportunities that assist in advancing their self-determined goals, and to reject those that do not. Current shifts in settler colonial government policy may offer further opportunities that nations can leverage from to support their strategising for self-governing.

**Keywords** Aboriginal and Torres Strait Islander · Indigenous nation building · Indigenous policy · Self-determination · Settler colonialism

---

M. Jorgensen (✉)
Native Nations Institute, University of Arizona, Tucson, AZ, USA
e-mail: mjorgens@arizona.edu

A. Vivian · A. Compton
Jumbunna Institute for Indigenous Education and Research, University of Technology Sydney, Ultimo, NSW, Australia

D. Murray
University of Technology Sydney, Ultimo, NSW, Australia

D. Evans
Charles Sturt University, Wagga Wagga, NSW, Australia

J. Gertz
Nura Gili, University of New South Wales, Kensington, NSW, Australia

 
N. Moodie and S. Maddison (eds.), *Public Policy and Indigenous Futures*, Indigenous-Settler Relations in Australia and the World 4,
https://doi.org/10.1007/978-981-19-9319-0_8

## 8.1 Introduction

If current rhetoric from federal and state governments is to be believed, Australia is potentially at a turning point in terms of relations among Aboriginal and Torres Strait Islander Peoples and settler colonial governments.[1] In response to decades of advocacy by Aboriginal and Torres Strait Islander people, the governments of Victoria, Queensland, and the Northern Territory have committed to creating frameworks for the negotiation of treaties that, by definition, would require those states to engage Indigenous Peoples as self-governing polities.[2] In September 2020, the Victorian Government also announced the availability of the Traditional Owner Nation-building Support Package, providing resources to Traditional Owners in Victoria for nation-building and treaty readiness activities (Victorian Government, 2020). The Federal Government similarly signalled a shift in its approach, recognising, as Prime Minister Morrison stated in February 2019, that the 2008 Closing the Gap policy had been 'set up to fail' because it was a 'top down approach' (Morrison, 2019). Efforts to reform followed, leading the Coalition of Aboriginal and Torres Strait Islander Peak Organisations to express their support for the 'genuine partnership' that led to the 2020 National Agreement on Closing the Gap (Australian Government, 2020).

On its face, such rhetoric suggests a new era. Looking deeper, settler colonial governments in Australia do not have a strong record of supporting self-determination for Aboriginal and Torres Strait Islander Peoples. It is also a mistake to assume that the settler state is presently able to lead the transition required to achieve co-sovereignty and equitable relations.[3] Until now, and following a pattern replicated by settler colonies globally, the Australian state has acted to achieve the effective disappearance of Indigenous Peoples as distinct collectives with inherent sovereign rights (Veracini, 2018; Wolfe, 2006)—initially through violence and dispossession; then through policies of assimilation, child removal, and relocation; and most recently through policies of normalisation and 'mainstreaming'. Moreover, because settler colonialism's relentless project is to concretise the institutional subordination of Indigenous Peoples, efforts to suppress Aboriginal and Torres Strait Islander nationhood are not now 'merely historical'. Rather, they persist in the fabric of settler colonial governance (Eatock, 2018; Strakosch, 2019; Vivian & Halloran, 2021).

Nonetheless and looking deeper still—beyond both the innovation and inertia of the settler colonial state—a movement toward Indigenous self-government already

[1] In this chapter, we use the term settler colonial governments to refer to federal, state and/or local governments to distinguish them from Aboriginal and Torres Strait Islander governments (discussed below).

[2] Treaties are understood to be legally binding agreements between sovereigns or governments. For example, the Victorian Government has defined a treaty as 'an agreement between states, nations or governments. This can include an agreement between Indigenous peoples and governments' (Victorian Government, 2021).

[3] We refer to equitable relations rather than equal relations because we acknowledge the prior and superior sovereignty of Aboriginal and Torres Strait Islander Peoples compared to that of the settler colonial state. In other words, we acknowledge Aboriginal and Torres Strait Islander sovereignty as having primacy.

is underway. Even without recognised legal personality under settler colonial law, Aboriginal and Torres Strait Islander collectives are opportunistically using settler colonial government policies, programs, and activities to further their own interests as self-defined, autonomous peoples who have never ceded sovereignty, and to exercise their rights to self-determination. We see Indigenous *nations* using settler colonial government policy and programs to suit their own purposes, and not necessarily for the purposes of said legislation or policy.[4] In doing so, we see *Indigenous nation building*.

This chapter considers the implications for Indigenous public policy of this moment in time—as defined by the orientation of settler colonial public policy—and of the possibilities demonstrated by Aboriginal and Torres Strait Islander communities' engagement in nation building. While the term 'Indigenous nation building' (INB) is used in a range of contexts in Australia, in our work as INB scholars, theorists, and practitioners, we define the term in line with its original usage and as reflected in the principles arising from decades of research by the Harvard Project on American Indian Economic Development (the Harvard Project) and its sister institution, the Native Nations Institute (NNI) at the University of Arizona. Centring this understanding of INB, we then provide three examples of Aboriginal polities in Australia undertaking nation building activities. The Gunditjmara People, Gugu Badhun Nation, and Wiradjuri Nation are all working to rebuild their governing foundations, strengthen their community governance, and, in so doing, advance nation-wide goals in an inherently hostile environment. In the final sections of this paper, we consider INB strategies in the context of Indigenous public policy—or more precisely, we consider INB strategies as the public policies of the Gunditjmara, Gugu Badhun, and Wiradjuri nations—and analyse how Aboriginal and Torres Strait Islander nation policies and settler colonial policies interact to establish 'now' as a propitious time for INB.

## 8.2 What Is Indigenous Nation Building?

The term 'Indigenous nation building' (INB) has existed at the edges of the academic and grey literature concerning Aboriginal and Torres Strait Islander community development since the early 2000s. Twenty years on, however, it has become much

---

[4] We use the term 'nation' to refer to Indigenous collectives that are, or seek to be, self-governing. While some Indigenous collectives prefer to use other terms to describe their legal and social personality, nation is the self-referential term chosen by the polities in Australia with whom we work, including the Gugu Badhun Nation, Gunditjmara People, Ngarrindjeri Nation, Nyungar Nation and Wiradjuri Nation. It provides the language for Indigenous political collectives to immediately express aspirations that transcend colonial state preoccupations such as more efficient service delivery. It reflects various groups' identities as Indigenous Peoples with specific responsibilities for Country, culture and constituency, and expresses their desire for multi-issue and cross-issue thinking. And, it connotes the sovereign status that Aboriginal and Torres Strait Islander Peoples repeatedly state they have never ceded.

more prominent. Whether due to its use in certain state and federal policies, to the traction it has achieved among Traditional Owners and native title holders, to the focus it provides for academic policy centres' interdisciplinary work, or to the sense-making it offers to grassroots community activists, the term has become nearly ubiquitous among scholars, practitioners, and policymakers working on Australian Indigenous governance concerns.

For better or worse, such widespread usage has spawned many strands of meaning (ATNS, 2020). On the upside, these variations may reflect the differing ambitions of Indigenous Peoples. On the downside, they risk of dilution of the term's original intent. Just as the term 'self-determination' at times is used in ways that constrain Indigenous Peoples' ambitions,[5] the current enthusiastic adoption of the term 'Indigenous nation building' could similarly strip from its meaning the basic principles underpinning INB, the decades of research that inform those principles, and the lessons for Indigenous nations that flow from them.

Our own usage reflects the roots and original meaning of INB. In scholarly literature, the first uses of 'nation building' as a reflexive activity of Indigenous nations, as opposed to something that is 'done to' Indigenous communities as part of nationalism-promoting activities of settler colonial states, were in the mid-late 1990s. According to Google Scholar (18.03.22), legal scholar Robert Porter's, 1996 article, 'Akwe:kon Forum: In Troubled Times A Vision of Nation Building', may be the first such source. While Porter does not offer a definition of the term, he nonetheless refers to the nexus of issues that have, over time, solidified as the principles of INB articulated by the Harvard Project. The second published use may be the 1997 article, 'Developing Nation Models for American Indian Nation Building: Educating the Young Leaders of Our Internal Developing Nations,' by Rosemarie Blanchard, an educational policy scholar. Blanchard draws an explicit parallel between the efforts of tribes and the efforts of developing nations worldwide 'which are building and rebuilding their political and economic institutions after having been conquered and occupied by nations with different political cultures than theirs' (1997, p. 425).

The term took off, however, with the 1998 article, 'Sovereignty and Nation Building: The Development Challenge in Indian Country Today,' by Harvard Project founders Stephen Cornell and Joseph Kalt. Like Porter and Blanchard, they describe the essence of INB, although their descriptions have a more educative flavour. For example, they (1998, p. 187) state:

> The Indian nations of the United States face a rare opportunity...This opportunity is a political and organizational one. It is a chance to rethink, restructure, reorganize—a chance not to start a business or exploit an economic niche but to substantially reshape the future. It is the opportunity for nation-building.

[5] It is unlikely that Indigenous Peoples' advocacy spanning more than 20 years at the United Nations for the right to self-determination to be included in the UN Declaration on the Rights of Indigenous Peoples was to be able to self-manage settler colonial government services or have the right to be consulted about policy. Yet, this is how 'Indigenous self-determination' often is framed by settler colonial governments.

A page later they are even more explicit, describing Native nation building as the process in which Indigenous Peoples 're-envision their futures and rebuild their governments and their economic strategies so as to realize those futures' (Cornell & Kalt, 1998, p. 188).

Cornell and Kalt's seminal exposition was backed by a decade of work from the Harvard Project. Founded in the late 1980s to address questions such as, 'What was enabling some Indian nations to break away from the overall pattern of seemingly intractable poverty? What were the conditions for sustained economic development on American Indian reservations?' (Cornell, 2005, p. 207); the Harvard Project already had produced an impressive catalogue of economic and policy research. This body of work pointed to the importance of Indigenous nation self-determination and self-government[6] and, consequently, to the importance of activities that Indigenous nations themselves undertake to make their governing institutions more sovereign and more effective (see, for e.g., Cornell & Kalt, 1991, 1992, 1993, 1995, 1997, 2007; Dixon et al., 1998; Krepps, 1992; Krepps & Caves, 1994).

Subsequent publications by Harvard Project and NNI affiliates repeat the themes of this original usage. For example, Jorgensen (2007, p. xii) defines Indigenous nation building as 'the processes by which a Native nation enhances its own foundational capacity for effective self-governance and for self-determined community and economic development'. Henson et al. (2020, p. 2) similarly explain that it is the process of 'building and rebuilding the legal, political, and social institutions that undergird the realization of community core values and the successful pursuit of community-determined goals'.

One might assume that because INB theory and research originated in North America, where Indigenous nations' legal status, land title, and rights differ from the rights and status of Aboriginal and Torres Strait Islander people and Peoples; the North American findings, principles, and approaches do not apply in Australia. Despite the differences, this is not the case. As our own research and that of others attests,[7] there are Indigenous nations in Australia engaged in activities that constitute

[6] We prefer the term 'government', rather than 'governance' as it draws attention to the institutions, structure and processes that Indigenous nations design, create and use to self-govern. Additionally, the language of 'government' provides a way for non-Indigenous people, especially settler colonial governments, to comprehend the scope of responsibility and authority that Indigenous self-governing Peoples seek to exercise (see Cornell, 2015; Vivian et al., 2018).

[7] In our research partnership with the Gunditjmara People and Ngarrindjeri Nation, and later individuals and groups from the Wiradjuri, Gugu Badhun and Nyungar Nations, we investigated the relevance of the Harvard Project's principles in Australia (the findings of which are partly detailed in this chapter). We found that the principles arising from North America resonated with our observations, and were useful to Aboriginal and Torres Strait Islander nations—in short, that the findings of the Harvard Project were highly relevant. Research projects have included Australian Research Council projects: 'Negotiating a space in the nation: the case of Ngarrindjeri' (DP1094869); 'Indigenous nationhood in the absence of recognition: Self-governance insights and strategies from three Aboriginal communities' (LP140100376); and 'Prerequisite conditions for Indigenous nation self-government' (DP190102060). A book detailing our initial inquiries with the Ngarrindjeri and Gunditjmara nations is forthcoming (Behrendt et al., in press).

'real' governing, and in so doing are accomplishing self-determined, Indigenous nation-level goals.

Of course, due to the destructive and pervasive history of settler colonialism in Australia, many Indigenous nations do not currently have operational institutions of self-governance. Nonetheless, Aboriginal and Torres Strait Islander communities' nation building work is unlikely to start with the creation of governments. Before discussions about governmental forms can be fruitful, more fundamental questions—concerning who 'the people' that constitute the nation are, what the nation stands for, what the nation's citizens want their united voice to say and what they want to accomplish together (see, for e.g., Cornell, 2000, 2013)—must be answered. A framework arising from the Harvard Project and NNI's research in North America and the Jumbunna-NNI research collaboration in Australia describes the iterative processes—identifying as a nation, organising as a nation, and acting as a nation—that generally occur as Indigenous nations address these questions.[8] Often, we observe these processes as individuals transition from viewing themselves as community members to viewing themselves as nation citizens, and as decision-making bodies capable of performing as governing bodies/governments emerge. As new circumstances, new challenges, and new opportunities arise, the processes can repeat and build upon one another through feedback loops, and further strengthen the nation over time.

By providing this clarity about the meaning of INB and the principles and activities it embodies, we aim to assist Indigenous communities in understanding INB as a strategic and organisational logic that supports their work towards self-determination. We wish to emphasise INB as action undertaken by Indigenous Peoples, not settler colonial states: settler colonial governments should *facilitate* INB with funding and other resources, but the process is not theirs to own. We seek to defend the term from attempts to subvert its potency (whether deliberately or inadvertently). Finally, we hope to convey—to all and sundry, but perhaps most especially to non-Indigenous federal and state policymakers, program administrators in the 'Aboriginal sector', and other researchers—that INB consists of any actions that leaders and members of an Indigenous nation determine are necessary for strengthening the community's foundational capacity for collective self-determination and self-government.

[8] The framework builds in particular on the experience and theorising of American and Australian researchers: Professors Steve Cornell and Miriam Jorgensen from the Native Nations Institute (University of Arizona); Professor Daryle Rigney, Associate Professor Steve Hemming and Dr. Alison Vivian (University of Technology Sydney); and Dr. Mark McMillan. For a detailed discussion, see Cornell (2015).

## 8.3 Three Indigenous Nation Building Stories From Australia

The stories of the Gunditjmara People, the Gugu Badhun Nation, and the Wiradjuri Nation are fruitful examples of INB at work. These nations are in different stages of institutional development for self-government but are all applying nation building principles and strategically taking action to strengthen their nationhood. Their experiences emphasise not only the reality of what is possible in Australia but the positive results that follow from undertaking the approach to INB as originally defined.

### *8.3.1 The Gunditjmara People*[9]

Gunditjmara Country, in the southwest of what is now known as Victoria, is extraordinary. Fertile. Rich volcanic plains, rugged coastline, limestone caves, forests and rivers, underground aquifers, and geothermal energy provide the abundant resources that have supported the Gunditjmara People for millennia (Weir, 2009). Gunditjmara oral history includes what may be the oldest story on earth—that of an ancestral creation being who revealed himself as Budj Bim, the volcano formerly known as Mount Eccles (Barras, 2020). For at least 6,600 years (McNiven et al., 2012), Gunditjmara engineers have used Budj Bim's extensive lava flow to construct and maintain a vast aquaculture complex and large network of stone house villages for which they are nationally and internationally renowned (The Gunditjmara People with Wettenhall, 2010).

The Gunditjmara People's highest priority since invasion has been to return Gunditjmara Country to Gunditjmara control. As Gunditjmara man Damein Bell has stated, 'You don't own Country, you own the obligation to care for Country' (Rigney et al., 2021, p. 123). Whether it was waging a guerrilla campaign against settlers during the Eumerella Wars in the mid-1800s, refusing the Aboriginal Protection Board's repeated attempts throughout the late 19th to mid-twentieth centuries to relocate them away from Gunditjmara Country, or making numerous appeals to colonial and Victorian authorities to return the Lake Condah mission site to Gunditjmara ownership, the nation has acted to protect, gain, and maintain authority over its Country. In the early-1990s, however, the nation lost control of parts of Country it had fought hard to regain, an event that ultimately proved pivotal to its rebuilding efforts and substantial successes today.

A dispute between the Alcoa Corporation and the Gunditjmara People over Alcoa's intention to build a smelter in Portland on Gunditjmara Country, led to an out-of-court settlement with the Victorian Government, which had courted Alcoa

[9] 'The Gunditjmara People' is the name chosen for the social and political collective—the nation—of Gunditjmara clans. Our account of the Gunditjmara People's nation building work emerges from our ongoing partnership with the Gunditj Mirring Traditional Owners Aboriginal Corporation (GMTOAC), including multiple ARC projects. It is provided with GMTOAC's permission.

to build the smelter (Horne, 2009). The settlement returned the Lake Condah mission site and several farms to Gunditjmara control and provided a compensation package. Unfortunately, the organisation created to manage the settlement lands and resources, the Kerrup-Jmara Elders Aboriginal Corporation, had inadequate management experience. Its practices resulted in a debt so substantial that the only recourse was to sell some of the land, with the remaining parcels removed from community control by the Victorian Government and put into trust. Perceptions of incompetence and recrimination from inside and outside the community compounded the Gunditjmara People's sense of loss at this turn of events.

While these outcomes were devastating, they also triggered deep reflection. The Gunditjmara People contemplated what had gone wrong, comprehensively analysing what it needed to do to restore its position as decision-maker so as to fulfil its responsibilities for Gunditjmara Country. This self-reflection and analysis led the community to adopt two transformative approaches to community governance. The first was a community-wide, centralised decision-making process developed on the principles of transparency, accountability, self-determination, and free, prior and informed consent. The process included a monthly community meeting known as the Gunditjmara Full Group that ultimately provided the means for the nation to fulfil significant collective aspirations. The second transformative approach was a conscious strategic orientation applied to the nation as a whole, supported by careful planning and patience.

Almost 20 years of monthly community meetings, and even more years of strategic planning and 'stick to it' capacity building; created stability, predictability and pure *ability*. In other words, organising as a nation made it possible for the Gunditjmara People to act as a nation. An era of notable strategic coherence followed, which further enabled the Gunditjmara People to pursue certain external opportunities that supported its nation priorities. Amongst others, successes achieved during this time include:

- two determinations of native title by consent;
- registration as a Registered Aboriginal Party with responsibility for protecting Gunditjmara cultural heritage;
- placement of the Budj Bim cultural landscape on the National Heritage list (the first listing of a site specifically for its Indigenous values);
- co-management of the Budj Bim National Park;
- reconstruction of an ancestral weir and reestablishment (through reflooding) of Lake Condah as a wetlands area;
- the return and reburial on Country of ancestors once housed in museums abroad;
- creation of the Gunditj Mirring Traditional Owner Corporation business centre and a museum grade keeping place;
- initiation of sustainable ecotourism plans despite disruption from COVID-19;
- recommencement of commercial eel aquaculture (to be operational by 2025); and
- inscription of the Budj Bim Cultural Landscape on the United Nations Educational, Scientific and Cultural Organization's World Heritage list.

Significantly, the Gunditjmara People's strategic planning is a nation-wide undertaking coordinated through a centralised decision-making process.[10] While individual community organisations and corporations create their own strategic plans to fulfil their statutory responsibilities and their rule books, the Gunditjmara People also undertake community planning based on well known, broadly understood, and widely accepted community aspirations. As Uncle Denis Rose, who led the World Heritage bid explains, a person can act with confidence knowing that the community has given them authority (2018, pers. comm.). Each comprehensive plan documents various strands of action, to be undertaken by different individuals, groups, or organisations, that in their own distinctive ways contribute to the overall imperative of reclaiming authority over Gunditjmara Country.

### *8.3.2 Gugu Badhun Nation*

The Gugu Badhun Nation, or the People of the Valley of Lagoons, are the Traditional Owners of the upper Burdekin region in what is now known as north Queensland. Having faced many of the same oppressions and subjugations as other Aboriginal and Torres Strait Islander nations—the loss of ancestral homelands to pastoralists, the appropriation of subsurface land by Australian governments and mining companies, the dislocation of population to urban centres across the region and farther, and diminution of cultural knowledge and connection—the Gugu Badhun have, especially in recent years, responded with a strategy focused on the re-creation of institutions for community decision-making and the exercise of community-level responsibilities.

Whilst there are historical examples of INB in Gugu Badhun's modern history, much of this work began in 2012, following the Gugu Badhun native title consent determination. Frustrated with the community's history of land loss and unsatisfactory agreements with resource appropriators, community members reasoned that the best path forward was to develop institutions capable of engaging with Australian settler colonial governments and other outside entities as equals. In other words, the Gugu Badhun sought to challenge the characterisation of Indigenous nations as 'stakeholders'. Instead, they aspired to create conditions in which engagement with settler colonial governments occurred on a government-to-government basis and engagement with private sector actors occurred on the basis of their status as senior rights holders. In order to accomplish these aspirations, Gugu Badhun leaders knew that they needed something more than what could be offered by a run-of-the mill Prescribed Body Corporate (PBC) designed to manage and protect their native title rights and interests. Instead, they needed an organisational structure that

[10] Until the establishment of GMTOAC in 2005, the centralised and coordinating body was Winda-Mara Aboriginal Corporation, followed by the Gunditjmara Full Group between 2002 and 2018. Since 2018 GMTOAC has again undertaken this role.

could accommodate, represent, make policy decisions about, and speak for all Gugu Badhun issues.

Writing elsewhere, Gugu Badhun citizen and scholar Janine Gertz (2021) makes exactly this point, referring to PBCs as a 'state constructed apparatus'. According to Gertz (2021), 'Whilst local level native title corporations hold significant prospect in exercising broader cultural, social, and economic aspirations of an Indigenous nation, the primary design and function of these entities is for the corporate governance and administration of native title'.

With this vision in mind, the Gugu Badhun Nation has embarked on a strategic pathway that involves the 'quadruple bottom-line issues of community, culture, economy and environment (country)' (GBAC, 2019). This process is jointly led by the Board of the Gugu Badhun Aboriginal Corporation (the Gugu Badhun PBC) and nation Elders, and has included the creation of a community development and economic development strategy; efforts to educate citizens about what it would mean to have a Gugu Badhun government; activities to reinforce Gugu Badhun identity and 'civic pride' among Gugu Badhun families and family members; and conversations about the appropriate role of a native title PBC within a broader governing structure.

Recent developments on Gugu Badhun Country have continued to emphasise the importance of having a having a Gugu Badhun government that can comprehensively manage an array of 'nationwide' issues into the future. Members of the Gugu Badhun Aboriginal Corporation Board and others (including Elders) have been involved in negotiations with the Australian Federal Government concerning irreversible development on Gugu Badhun Country. While the negotiators are acutely aware that the negotiations and the terms they accept will affect Gugu Badhun People for generations to come, there is no possibility of deferring negotiations, or waiting until the balance of power shifts. The immediacy of the negotiations and impact on all Gugu Badhun People, not just PBC members, has reinforced the need for other governing structures and processes that better accommodate a 'nation perspective.' Regardless, the fact that Gugu Badhun is engaged in nation-to-nation negotiations with the Australian federal government is evidence of the success of its nation building work to date. As it continues this work, it is likely that Gugu Badhun nation will further protect, manage, and expand its authority to advance its national aspirations.

### *8.3.3 Wiradjuri Nation*

The Wiradjuri Nation, also known as the People of the Three Rivers,[11] is one of the largest Aboriginal nations in Australia. The nation has an estimated population of 28,000–30,000 citizens, with its Country extending over vast areas of what is now known as central NSW. Because some of the most severe assimilationist and homogenising policies of the early colonial era were targeted at this region,

[11] The three rivers are the Wambool (also known as the Macquarie), the Kalare (also known as the Lachlan) and the Murrumbidjeri (also known as the Murrumbidgee).

Aboriginal communities in towns across Wiradjuri Country today frequently are composed of people who either have chosen to, or been forced to, move far from their own Country, creating highly heterogeneous communities. As a result, Wiradjuri Traditional Owners often are a minority in their 'home' Aboriginal communities. Moreover, these same push-and-pull forces have led Wiradjuri people to settle across Australia, so that many more Wiradjuri People live off Country than on.

These facts have created a situation in which Wiradjuri People's connections to Country are sometimes attenuated, knowledge of Wiradjuri culture and heritage is highly variegated, ideas of what it means to be a Wiradjuri citizen exist across a wide spectrum, many Wiradjuri people feel more allegiance to their town or region than to their nation, and the Wiradjuri Nation's ability to operate as political collective is as-yet limited. As Uncle Jimmy Ingram explained at a Wiradjuri nation building gathering in 2017, 'Until we bring Wiradjuri people together, we won't get anywhere' (quoted in Murray & Evans, 2021, p. 180). Later in the same conversation, he expanded on the point, noting 'now is not the time to rebuild a government, now is the time to rebuild connections to one another as one nation' (pers. comm. 2017). Wiradjuri nation builders therefore are focused on the element of identifying as a nation. They are taking a cultural development approach to nation-building, working to build Wiradjuri connections to people and Country through the sharing and teaching of Wiradjuri language, knowledges, and ways of being (see Murray & Evans, 2021).

Perhaps the most notable Wiradjuri nation building effort has occurred through the leadership of the Wiradjuri Council of Elders. In the late 2000s the Council became frustrated that Charles Sturt University (CSU), a major regional university with six campuses on Wiradjuri Country, was insufficiently engaged with the Wiradjuri community. The Wiradjuri Council of Elders' position was that the support CSU provided to Aboriginal people generally was not enough. Instead, CSU needed to show its commitment to the Wiradjuri Nation as the Traditional Owners of Country from which the university as an institution—and all its faculty, staff, students, and alumni—benefited.

This nationhood-grounded advocacy by the Council of Elders paid off: CSU agreed to help create and fund a Graduate Certificate in Wiradjuri Language, Culture and Heritage (GCWLCH).[12] Devised primarily by Uncle Stan Grant and Aunty Flo (Florence) Grant under instruction from the Council (with the assistance of Wiradjuri nation builders and non-Indigenous allies), the GCWLCH's four subjects—deep language, deep culture, Wiradjuri nation-building, and community service—each have Wiradjuri cultural concepts at their core. Further, the GCWLCH is taught by Wiradjuri people according to the concept of Yindyamarra (respect), which embodies Wiradjuri sovereignty and self-determination and privileges Wiradjuri pedagogical practices, including multi-generational engagement, service learning and connection to Country.

Serendipitously, during the development phase of the GCWLCH, two drivers of the program, chapter authors Donna Murray (Wiradjuri) and Debra Evans

[12] The GCWLCH is open to both Wiradjuri and non-Wiradjuri students.

(Barkindji), travelled to the University of Arizona to participate in an intensive Indigenous governance program. They learned alongside Indigenous people from around the world who also were working to strengthen and rebuild their nations, and they visited four Native nations to witness how those nations' governing systems operate. These experiences resulted in two transformative learnings. First, Murray and Evans gained access to research findings that proved what they had suspected from decades of frustrating community development work. They confirmed that so long as settler colonial government policy and programs took precedence over the cultural, social, economic, and political aspirations of their communities, progress would be thwarted; and that 'self-governance' as defined, designed, and implemented by Indigenous communities themselves was the missing ingredient. Second, they acquired the evidence and language necessary to articulate what Wiradjuri Elders had been telling them but that they had not fully understood. In particular, Wiradjuri spiritual leader and cultural Elder Wongamarr (also known as Pastor Cec Grant) had repeatedly said: 'We have to start thinking in a Wiradjuri way and we have to heal our nation before we can heal our people' (Murray & Evans, 2021, p. 165).[13] These learnings are deeply embedded in the GCWLCH. The goal has been to shift the conversation from one in which Wiradjuri cultural identity is 'a problem', or at best 'just words', to one in which Wiradjuri culture and collective being is the foundation on which the nation and its people can rebuild a positive and vibrant future.

Offered for the first time in 2014, the certificate program has become a resounding success. Growth has been exponential, with demand outstripping available placements at a ratio of 3 to 1. Completion rates challenge the widely held view of failure and under-performance in Aboriginal and Torres Strait Islander education. Each year, graduates return to mentor the current cohort and to renew their learning about Wiradjuri language, culture, heritage, and contemporary nationhood. GCWLCH graduates have pursed different employment opportunities, commenced additional study, and started their own efforts to teach Wiradjuri language, all as part of their efforts to fulfil newly understood responsibilities.

In sum, the GCWLCH has created a network of Wiradjuri nation builders. As Murray and Evans (pers. comm. 2019) explain, 'Somehow people are changed: their thinking changes, conversations change, people start to do things differently and importantly, they now have a very different set of relationships and connections in their lives. Certainly, they have a different lens for deciding when and where to focus their time and energy.' Similarly, Jonathan Jones, a world-renowned Wiradjuri artist and GCWLCH graduate, has said, 'Set within the Wiradjuri framework, you are immersed in our ways of being and understanding. …This course brings us together as a nation with one heart' (Charles Sturt University, 2019).

Significant developments have followed this network of Wiradjuri nation builders. Since 2020, some Wiradjuri have been undertaking community genealogies and histories, working to reclaim Wiradjuri clan group names, structures, and identities.

[13] Wongamarr was also fierce in denouncing what he called 'Aboriginalisation,' which he saw as a form of brainwashing.

Using Wiradjuri language, community members have been identifying clan boundaries based on the Wambool, Kalare, and the Murrumbidjeri rivers, with a view to initiating conversations about a contemporary governance system. Preliminary planning has focused on ensuring that Wiradjuri law and knowledge is not only successfully incorporated, but foundational to, any community governance structure created.

## 8.4 Indigenous Nation Building and Aboriginal and Torres Strait Islander Nation Policy

The Gunditjmara People, Gugu Badhun Nation, and Wiradjuri Nation are demonstrably nation building. They are working to strengthen their capacities to govern their communities (now or in the future) with legitimacy and efficacy and are achieving community goals on their communities' terms.

These achievements stand in stark contrast to the logic of settler colonisation, which requires the displacement of Indigenous Peoples, the erasure of cultural practices and the 'absorption', 'assimilation', and 'amalgamation' of Indigenous Peoples into the wider population (Veracini, 2011, p. 2). The logic of settler colonialism is evident in Australian legal and political institutions that reflect the myth of legal centralism, which claims there can be only one law for all and only one set of institutions to administer the law (Griffiths, 1986); through a national narrative that asserts a single, indivisible Australian sovereignty and that all citizens belong to a monocultural society and possess the same civil, political, social, cultural and economic rights (Moreton-Robinson, 2007); and through policies of normalisation, whereby Indigenous people are encouraged, coerced or forced to reproduce lifestyles of the broader Australian community (Eatock, 2018; Sullivan, 2013).

What is more, the logic creates a tendency to focus solely on settler colonial state actions and the impact of those actions upon Aboriginal and Torres Strait Islander people; and to view the reform of settler colonial practices, programs, and policies as the only possible evidence of 'progress' in Indigenous affairs in Australia. Significantly, this restricted view ignores (or at the very least discounts) the self-determined actions of Aboriginal and Torres Strait Islander Peoples. As Rowse observes, 'every instance of Indigenous agency is under suspicion of being "state-conceded"' (Rowse, 2014, p. 300). This in turn makes it difficult to see the actions of Aboriginal and Torres Strait Islander Peoples as the actions of nations that are responding, on their own terms, to a hostile legal-political-economic environment.

In this light, the INB stories recounted above reveal an important means by which Aboriginal and Torres Strait Islander Peoples have prevailed over settler colonialism: they have prevailed through their actions as Indigenous nations, and *through their policies*. The Gunditjmara, Gugu Badhun, and Wiradjuri nations' persistent and consistent assertions that they never ceded sovereignty, their continued claim to status as distinct and autonomous peoples with an inherent right to self-determination, and their unrelenting advocacy to exercise inherent rights as Peoples are overarching policy strategies in support of nationhood—as are their INB strategies and goals.

Looking across the three case examples, a common strategy used by the Gunditjmara People, Gugu Badhun Nation, and Wiradjuri Nation to achieve their INB aspirations is to use—but not rely on—settler colonial policies. This is a subtle but elemental distinction: the nations look for spaces to exercise self-determination and expand jurisdiction, seize autonomy, and push past settler colonial bureaucracy, and thereby treat settler colonial policy and programs as a means, rather than an end. Succinctly, these Indigenous nations' policy is to use settler colonial policy for their own purposes.[14]

The Gunditjmara People's decision to lodge an application for native title determination is one such example. While the resultant consent determinations ultimately have been sources of pride and platforms for collective self-determination, the nation did not view the native title process as a means of validating the Gunditjmara People's authority or of proving the collective's legitimacy to the settler colonial state. Instead, the community's decision to apply to the settler state legal system to be acknowledged as Traditional Owners was one of the Gunditjmara People's various strategies to reclaim control of its Country. In other words, the Gunditjmara People's participation in this settler colonial scheme was not an aspiration in and of itself; the Gunditjmara People deliberately sought it out as a way to use settler colonial resources and opportunities to achieve its nation building goals.

Similarly, the Gunditjmara People has rejected state and Commonwealth policies and decisions that, regardless of how beneficial they were intended to be from a settler colonial perspective, did not progress the nation toward its aspirations as it has determined them. The community's 2015 rejection of the financial incentives offered through the Victorian Traditional Owners Settlement Act is one example: the Gunditjmara People turned down the offer in part because it decreased its authority on Country. The Gunditjmara People's bid for World Heritage status is another: had the Australian Government decided not to nominate the Budj Bim Cultural Landscape for World Heritage Status, the Gunditjmara People was willing to reject the settler colonial government's decision and put forth a nomination based on its own nationhood, again with the rationale that gaining such status would significantly increase Gunditjmara control over Gunditjmara Country.

In the same vein, the Gugu Badhun Nation is highly strategic in negotiations with the Australian federal government. Alongside protecting particular Gugu Badhun interests, negotiators are using the process to further settler recognition of the Gugu Badhun nation as a distinct, sovereign polity with responsibility for Gugu Badhun Country.

Wiradjuri Elders also have leveraged from available opportunities to advance Wiradjuri nation building. Building on CSU's commitment to positive relationships

[14] A critical benefit of this approach is that community development gains are less likely to crumble when settler colonial policy changes. The reason for such sustainability is that strategic participation differs from participation through default or desperation. With self-determined choice, an Indigenous nation is more likely to 'own' the change.

with the Aboriginal and Torres Strait Islander people, Wiradjuri Elders pursued and realised specific and non-negotiable goals that entailed service to, and deference to, Wiradjuri people as Traditional Owners. Although the GCWLCH is a qualification offered by CSU, it is under Wiradjuri control, deeply embeds Wiradjuri principles and values in teaching and learning and fulfils Wiradjuri aspirations.

## 8.5 Concluding Thoughts

In their pivotal commentary 'Sovereignty and Nation Building', Cornell and Kalt reflect on a particularly hopeful period for INB in the United States (1998, p. 188):

> [American] Indian nations should determine their own futures and determined Indian nations can still do so. But shaping those futures will require not simply the assertion of sovereignty—a claim to rights and powers—it will require the effective exercise of that sovereignty. The task tribes face is to use the power they have to build viable nations before the opportunity slips away. This is the major challenge facing Indian country today. It also is the key to solving the seemingly intractable problem of reservation poverty. Sovereignty, nation building, and economic development go hand in hand.

Their reflection directs attention to two important questions for Indigenous Peoples: why should they engage in nation building at all, and why should they engage in nation building *now*?

We began this chapter by reprising the Harvard Project's and NNI's evidence-based meaning of the term 'Indigenous nation building', a recapitulation that points to the intertwined nature of the assertion of nationhood, the creation and strengthening of governing institutions and the sustainability and reinvigoration of Aboriginal and Torres Strait Islander societies. INB is not simply a means by which Aboriginal and Torres Strait Islander people are assured greater civil or human rights protections or fairer access to services, resources, or lands. Nor, as Blanchard writes, is it 'just patching up a broken past … [or] just "hanging in there" for a while longer' (1997, p. 430). It is about communities that have existed since time immemorial creating the conditions for an auspicious future, a time without end. INB as a practice is imbued with promise for Indigenous Peoples; *that* is the reason to pursue it.

We continue by recounting the Gunditjmara People's, Gugu Badhun Nation's, and Wiradjuri Nation's strategic efforts to assert authority, expand jurisdiction and fortify their collective identities. They variously have sought to establish or strengthen their decision-making mechanisms and to build on-Country relationships with settler colonial governments, business, industry, regional decision-makers, landowners, and interest groups that reinforce their claims to sovereign status. Each is undertaking nation building actions that meet the dictionary definition of public policymaking: 'the act or process of setting and directing the course of action to be pursued by a government'.[15] In particular, each is leveraging available external opportunities to further nationhood goals.

[15] See the definition of 'policymaking' in Ballentine's Law Dictionary (2021). Crawford's description of 'Aboriginal public law' as the 'body of rules, traditions or understandings' that define the

It is such stories and analyses that underline our impression that just as tribes in the United States were during the 1990s, Aboriginal and Torres Strait Islander nations in Australia may be in the midst of an encouraging time for INB. State governments' commitments to treaty conversations and self-determination options may provide Aboriginal and Torres Strait Islander nations with footholds from which to progress their own their strategies for self-determination and self-governance. In other words, we are cautiously hopeful about these shifts in the rhetoric and policies of settler colonial governments. Given the examples of the Gunditjmara, Gugu Badhun, and Wiradjuri nations, positive outcomes are possible: especially if they frame these settler colonial initiatives as strategic opportunities and pursue complementary policies of their own, many more Aboriginal and Torres Strait Islander nations may be able to make significant jurisdictional gains and build towards greater collective self-determination. So 'yes, the time is now' for INB.

And yet, we are realistically wary not only of the proposed measures' capacity to deliver the fundamental transformation of relations needed for Aboriginal and Torres Strait Islander Peoples to flourish (see Vivian & Halloran, 2021) but also of settler colonial governments' faithfulness. Federal and state Australian governments are not well-known for the consistency or longevity of policies that support self-determination, and each about-face has the potential to leave Aboriginal and Torres Strait Islander nations in a worse position than before. To iterate a point made above, *relying* on settler colonial governments for nation building opportunities can be ill conceived, and potentially even damaging.

It is also a stinting approach to nation building. Most fundamentally, the Gunditjmara, Gugu Badhun, and Wiradjuri examples demonstrate that no settler colonial permissions are needed, and no particular settler colonial investments (while undoubtably useful) are required, for INB to proceed. The considerable benefits to be obtained from self-government and self-determination make it both reasonable and imperative for Aboriginal and Torres Strait Islander communities to pursue the processes of Indigenous nationhood as persistently as they can. Further, it is only by adopting this persistent approach—in which Aboriginal and Torres Strait Islander communities are always working to strengthen their collective identities and their capacities for self-governance—that INB progress can endure if (or perhaps inevitably, when) Australian government policies change again.

From this perspective, the time for INB is *always* now.

## References

Agreement, Treaties and Negotiated Settlements (ATNS). (2020). *Nation building in Australia*, viewed 5 January 2022, https://www.atns.net.au/nation-building-landing

Australian Government (2020). *National Agreement on Closing the Gap*. https://www.closingthegap.gov.au/national-agreement

---

Indigenous policy and provide the authority to members of that community to act as a political collective is a related, and precursor, concept (Crawford, 1989, p. 397).

Ballentine's Law Dictionary. (2021). *Policymaking*. https://ballentine.en-academic.com/29131/policymaking. Accessed 31 Jan 2022.

Barras, C. (2020). Is an aboriginal tale of an ancient volcano the oldest story ever told? *Science*, viewed 21 December 2021, https://www.sciencemag.org/news/2020/02/aboriginal-tale-ancient-volcano-oldest-story-ever-told; https://doi.org/10.1126/science.abb2656

Blanchard, R. A. (1997). Developing nation models for American Indian nation building: Educating the young leaders of our internal developing nations. *Social Science Journal, 34*(4), 423–434.

Charles Sturt University. (2019). *Charles Sturt University and Wiradjuri Elders win global citizenship*, viewed 10 December 2021, https://news.csu.edu.au/latest-news/charles-sturt-university-and-wiradjuri-elders-win-global-citizenship

Cornell, S. (2000). That's the story of our life. In P. Spickard & W. J. Burroughs (Eds.), *We are a people: Narrative and multiplicity in constructing ethnic identity* (pp. 41–51). Temple University Press.

Cornell, S. (2005). Indigenous peoples, poverty, and self-determination in Australia, New Zealand, Canada, and the United States. In R. Eversole, J. A. McNeish, & A. D. Cimadamore (Eds.), *Indigenous peoples and poverty: An international perspective* (pp. 199–225). Zed Books.

Cornell, S. (2013). Reconstituting native nations: Colonial boundaries and institutional innovation in Canada, Australia, and the United States. In R. Walker, T. Jojola, & D. Natcher (Eds.), *Reclaiming indigenous planning* (pp. 35–59). McGill-Queens University Press.

Cornell, S. (2015). Processes of native nationhood: The indigenous politics of self-government. *The International Indigenous Policy Journal, 6*(4), article 4.

Cornell, S., & Kalt, J. P. (1991, February). Where's the glue? Institutional bases of American Indian economic development. *Paper presented at the December 1990 National Bureau of Economic Research conference on political economy* (Revised).

Cornell, S., & Kalt, J. P. (1992). Reloading the dice: Improving the chances for economic development on American Indian reservations. In S. Cornell & J. P. Kalt (Eds.), *What can tribes do? Strategies and institutions in American Indian economic development* (pp. 1–59). University of California.

Cornell, S., & Kalt, J. P. (1993). *Culture as explanation in racial and ethnic inequality: American Indians, reservation poverty, and collective action* (Project Report Series No. 93-2). Harvard Project on American Indian Economic Development, John F. Kennedy School of Government, Harvard University.

Cornell, S., & Kalt, J. P. (1995). Where does economic development really come from? Constitutional rule among the contemporary Sioux and Apache. *Economic Inquiry, 33*(3), 402–426.

Cornell, S., & Kalt, J. P. (1997). Successful economic development and heterogeneity of governmental form on American Indian reservations. In M. S. Grindle (Ed.), *Getting good government: Capacity building in the public sectors of developing countries* (pp. 257–296). Harvard Institute for International Development.

Cornell, S., & Kalt, J. P. (1998). Sovereignty and nation building: The development challenge in Indian country today. *American Indian Culture and Research Journal, 23*(3), 187–214.

Cornell, S., & Kalt, J. P. (2007). Two approaches to the development of native nations: One works, the other doesn't. In M. Jorgensen (Ed.), *Rebuilding native nations: Strategies for governance and development* (pp. 3–33). University of Arizona Press.

Crawford, J. (1989). The aboriginal legal heritage: Aboriginal public law and the treaty proposal. *The Australian Law Journal, 63*, 392–403.

Dixon, M., Shelton, B. L., Roubideaux, Y., Mather, D., & Smith, C. M. (1998). *Tribal perspectives on Indian self-determination and self-governance in health care management*. National Indian Health Board.

Eatock, C. (2018). Resisting the ascendancy of an emboldened colonialism. In D. Howard-Wagner, M. Bargh, & I. Altamirano-Jimenez (Eds.), *The neoliberal state, recognition and indigenous rights* (pp. 59–75). Centre for Aboriginal Economic Policy Research (CAEPR) Research Monograph No. 40, Australian National University.

Gugu Badhun Aboriginal Corporation (GBAC). (2019). *Gugu Badhun aboriginal corporation strategic plan 2019–2024: Summary*. Gugu Badhun Aboriginal Corporation, on file with authors.

Gertz, J. (2021, November 29–December 1). Treaties, agreements and other constructive arrangements, between indigenous peoples and States, including peace accords and reconciliation initiatives, and their constitutional recognition. *UN Seminar of the Expert Mechanism on the Rights of Indigenous Peoples*. https://www.ohchr.org/EN/Issues/IPeoples/EMRIP/Pages/EMRIP-Seminar-Treaties.aspx. Accessed 28 Jan 2022.

Griffiths, J. (1986). What is legal pluralism? *The Journal of Legal Pluralism and Unofficial Law, 18*(24), 1–55.

Henson, E. C., Hill, M. M., Jorgensen, M. R., & Kalt, J. P. (2020). Emerging stronger than before: Guidelines for the federal role in American Indian and Alaska native tribes' recovery from the COVID-19 pandemic. *COVID-19 Response and Recovery Policy Brief Series* (Policy Brief No. 4). Harvard Project on American Indian Economic Development, and Native Nations Institute, The University of Arizona, viewed 21 December 2021, https://ash.harvard.edu/files/ash/files/policy_brief_4-federal_policy_24july2020_final_for_dist.pdf?m=1595612546

Horne, C. (2009). The history of the Alcoa aluminium smelter in Portland. *Chain Reaction, 107*, 24–26.

Jorgensen, M. (2007). Editor's introduction. In M. Jorgensen (Ed.), *Rebuilding native nations: Strategies for governance and development* (pp. xi–xiv). University of Arizona Press.

Krepps, M. B. (1992). Can tribes manage their own resources? A study of American Indian forestry and the 638 program. In S. Cornell & J. P. Kalt (Eds.), *What can tribes do? Strategies and institutions in American Indian economic development* (pp. 179–203). American Indian Studies Center.

Krepps, M. B., & Caves, R. E. (1994). Bureaucrats and Indians: Principal-agent relations and efficient management of tribal forest resources. *Journal of Economic Behavior and Organization, 24*(2), 133–151.

McNiven, I. J., Crouch, J., Richards, R., Dolby, N., Jacobsen, G., & Corporation, G. M. T. O. A. (2012). Dating aboriginal stone-walled fishtraps at Lake Condah, southeast Australia. *Journal of Archaeological Science, 39*(2), 268–286.

Morrison, S. (2019). Statement to the house of representatives—Closing the gap 2019. *Prime Minister of Australia*, viewed 20 January 2020, https://www.pm.gov.au/media/statement-house-representatives-closing-gap-2019

Moreton-Robinson, A. (2007). Writing off indigenous sovereignty: The discourse of security and patriarchal white sovereignty. In A. Moreton-Robinson (Ed.), *Sovereign subjects: Indigenous sovereignty matters* (pp. 86–102). Allen and Unwin.

Murray, D., & Evans, D. (2021). Culturally centred, community led: Wiradjuri nation rebuilding through honouring the Wiradjuri Way. In D. Smith, A. Wighton, S. Cornell, & A. V. Delaney (Eds.), *Developing governance and governing development: International case studies of indigenous futures* (pp. 165–186). Rowman and Littlefield International.

Porter, R. (1996). Akwe:Kon Forum: In troubled times a vision of nation building. *Native Americas, 8*(4), 52–57.

Rigney, D., Bignall, S., Vivian, A., Hemming, S., Berg, S., & Bell, D. (2021). Treating treaty as a technology for indigenous nation building. In D. Smith, A. Wighton, S. Cornell, & A. V. Delaney (Eds.), *Developing governance and governing development international case studies of indigenous futures* (pp. 119–140). Rowman and Littlefield International.

Rowse, T. (2014). Indigenous heterogeneity. *Australian Historical Studies, 45*(3), 297–310.

Strakosch, E. (2019). The technical is political: Settler colonialism and the Australian Indigenous policy system. *Australian Journal of Political Science, 54*(1), 114–130.

Sullivan, P. (2013). Disenchantment, normalisation and public value: Taking the long view in Australian indigenous affairs. *Asia Pacific Journal of Anthropology, 14*(4), 353–369.

The Gunditjmara People with Wettenhall, G. (2010). *The people of Budj Bim. Engineers of aquaculture, builders of stone house settlements and warriors defending country*, em PRESS Publishing for the Gunditj Mirring Traditional Owners Aboriginal Corporation, Heywood.

Veracini, L. (2011). Introducing settler colonial studies. *Settler Colonial Studies, 1*(1), 1–12.
Veracini, L. (2018). Containment, elimination, settler colonialism. *Arena Journal, 51*(52), 18–39.
Victorian Government. (2020). Traditional owner nation-building support package. *First Peoples Relations*, viewed 19 December 2021, https://www.firstpeoplesrelations.vic.gov.au/nation-building
Victorian Government. (2021). Treaty in Victoria. *First peoples relations*, viewed 27 January 2022, https://www.firstpeoplesrelations.vic.gov.au/treaty
Vivian, A., & Halloran, M. J. (2021). Dynamics of the policy environment and trauma in relations between Aboriginal and Torres Strait Islander peoples and the settler colonial state. *Critical Social Policy*, online, https://doi.org/10.1177/02610183211065701
Vivian, A., Jorgensen, M., Reilly, A., McMillan, M., McRae, C., & McMinn, J. (2018). Indigenous self-government in the Australian Federation. *Australian Indigenous Law Review, 20*, 215–242.
Weir, J. K. (2009). *The Gunditjmara land justice story*. Australian Institute of Aboriginal and Torres Strait Islander Studies.
Wolfe, P. (2006). Settler colonialism and the elimination of the native. *Journal of Genocide Research, 8*(4), 387–409.

The manufacturer's authorised representative in the EU is Springer Nature Customer Service Centre GmbH, Europaplatz 3, 69115 Heidelberg, Germany. If you have any concerns regarding our products, please contact ProductSafety@springernature.com

Printed and bound by CPI Group (UK) Ltd, Croydon, CR0 4YY
30/09/2026
02225264-0010